OFFICIAL SQA PAST PAPERS
WITH ANSWERS

INTERMEDIATE 2

PHYSICS
2009-2013

Hodder Gibson Study Skills Advice – General   – page 3
Hodder Gibson Study Skills Advice – Intermediate 2 Physics   – page 5
2009 EXAM   – page 7
2010 EXAM   – page 29
2011 EXAM   – page 51
2012 EXAM   – page 73
2013 EXAM   – page 101
ANSWER SECTION   – page 129

HODDER
GIBSON
LEARN MORE

Hodder Gibson is grateful to the copyright holders, as credited on the final page of the Question Section, for permission to use their material. Every effort has been made to trace the copyright holders and to obtain their permission for the use of copyright material. Hodder Gibson will be happy to receive information allowing us to rectify any error or omission in future editions.

Hachette UK's policy is to use papers that are natural, renewable and recyclable products and made from wood grown in sustainable forests. The logging and manufacturing processes are expected to conform to the environmental regulations of the country of origin.

Orders: please contact Bookpoint Ltd, 130 Park Drive, Abingdon, Oxon OX14 4SE. Telephone: (44) 01235 827720. Fax: (44) 01235 400454.

Lines are open 9.00–5.00, Monday to Saturday, with a 24-hour message answering service. Visit our website at www.hoddereducation.co.uk. Hodder Gibson can be contacted direct on: Tel: 0141 848 1609; Fax: 0141 889 6315; email: hoddergibson@hodder.co.uk

This collection first published in 2013 by

Hodder Gibson, an imprint of Hodder Education,

An Hachette UK Company

2a Christie Street

Paisley PA1 1NB

{BrightRED   Hodder Gibson is grateful to Bright Red Publishing Ltd for collaborative work in preparation of this book and all
PUBLISHING   SQA Past Paper and National 5 Model Paper titles 2013.

Typeset by PDQ Digital Media Solutions Ltd, Bungay, Suffolk NR35 1BY

Printed in the UK

A catalogue record for this title is available from the British Library

ISBN 978-1-4718-0265-2

3 2 1

2014 2013

Introduction

Study Skills – what you need to know to pass exams!

Pause for thought

Many students might skip quickly through a page like this. After all, we all know how to revise. Do you really though?

Think about this:

"IF YOU ALWAYS DO WHAT YOU ALWAYS DO, YOU WILL ALWAYS GET WHAT YOU HAVE ALWAYS GOT."

Do you like the grades you get? Do you want to do better? If you get full marks in your assessment, then that's great! Change nothing! This section is just to help you get that little bit better than you already are.

There are two main parts to the advice on offer here. The first part highlights fairly obvious things but which are also very important. The second part makes suggestions about revision that you might not have thought about but which WILL help you.

Part 1

DOH! It's so obvious but …

Start revising in good time

Don't leave it until the last minute – this will make you panic.

Make a revision timetable that sets out work time AND play time.

Sleep and eat!

Obvious really, and very helpful. Avoid arguments or stressful things too – even games that wind you up. You need to be fit, awake and focused!

Know your place!

Make sure you know exactly **WHEN and WHERE** your exams are.

Know your enemy!

Make sure you know what to expect in the exam.

How is the paper structured?

How much time is there for each question?

What types of question are involved?

Which topics seem to come up time and time again?

Which topics are your strongest and which are your weakest?

Are all topics compulsory or are there choices?

Learn by DOING!

There is no substitute for past papers and practice papers – they are simply essential! Tackling this collection of papers and answers is exactly the right thing to be doing as your exams approach.

Part 2

People learn in different ways. Some like low light, some bright. Some like early morning, some like evening / night. Some prefer warm, some prefer cold. But everyone uses their BRAIN and the brain works when it is active. Passive learning – sitting gazing at notes – is the most INEFFICIENT way to learn anything. Below you will find tips and ideas for making your revision more effective and maybe even more enjoyable. What follows gets your brain active, and active learning works!

Activity 1 – Stop and review

Step 1

When you have done no more than 5 minutes of revision reading STOP!

Step 2

Write a heading in your own words which sums up the topic you have been revising.

Step 3

Write a summary of what you have revised in no more than two sentences. Don't fool yourself by saying, 'I know it but I cannot put it into words'. That just means you don't know it well enough. If you cannot write your summary, revise that section again, knowing that you must write a summary at the end of it. Many of you will have notebooks full of blue/black ink writing. Many of the pages will not be especially attractive or memorable so try to liven them up a bit with colour as you are reviewing and rewriting. **This is a great memory aid, and memory is the most important thing.**

Activity 2 — Use technology!

Why should everything be written down? Have you thought about 'mental' maps, diagrams, cartoons and colour to help you learn? And rather than write down notes, why not record your revision material?

What about having a text message revision session with friends? Keep in touch with them to find out how and what they are revising and share ideas and questions.

Why not make a video diary where you tell the camera what you are doing, what you think you have learned and what you still have to do? No one has to see or hear it but the process of having to organise your thoughts in a formal way to explain something is a very important learning practice.

Be sure to make use of electronic files. You could begin to summarise your class notes. Your typing might be slow but it will get faster and the typed notes will be easier to read than the scribbles in your class notes. Try to add different fonts and colours to make your work stand out. You can easily Google relevant pictures, cartoons and diagrams which you can copy and paste to make your work more attractive and **MEMORABLE**.

Activity 3 – This is it. Do this and you will know lots!

Step 1

In this task you must be very honest with yourself! Find the SQA syllabus for your subject (www.sqa.org.uk). Look at how it is broken down into main topics called MANDATORY knowledge. That means stuff you MUST know.

Step 2

BEFORE you do ANY revision on this topic, write a list of everything that you already know about the subject. It might be quite a long list but you only need to write it once. It shows you all the information that is already in your long-term memory so you know what parts you do not need to revise!

Step 3

Pick a chapter or section from your book or revision notes. Choose a fairly large section or a whole chapter to get the most out of this activity.

With a buddy, use Skype, Facetime, Twitter or any other communication you have, to play the game "If this is the answer, what is the question?". For example, if you are revising Geography and the answer you provide is "meander", your buddy would have to make up a question like "What is the word that describes a feature of a river where it flows slowly and bends often from side to side?".

Make up 10 "answers" based on the content of the chapter or section you are using. Give this to your buddy to solve while you solve theirs.

Step 4

Construct a wordsearch of at least 10 X 10 squares. You can make it as big as you like but keep it realistic. Work together with a group of friends. Many apps allow you to make wordsearch puzzles online. The words and phrases can go in any direction and phrases can be split. Your puzzle must only contain facts linked to the topic you are revising. Your task is to find 10 bits of information to hide in your puzzle but you must not repeat information that you used in Step 3. DO NOT show where the words are. Fill up empty squares with random letters. Remember to keep a note of where your answers are hidden but do not show your friends. When you have a complete puzzle, exchange it with a friend to solve each other's puzzle.

Step 5

Now make up 10 questions (not "answers" this time) based on the same chapter used in the previous two tasks. Again, you must find NEW information that you have not yet used. Now it's getting hard to find that new information! Again, give your questions to a friend to answer.

Step 6

As you have been doing the puzzles, your brain has been actively searching for new information. Now write a NEW LIST that contains only the new information you have discovered when doing the puzzles. Your new list is the one to look at repeatedly for short bursts over the next few days. Try to remember more and more of it without looking at it. After a few days, you should be able to add words from your second list to your first list as you increase the information in your long-term memory.

FINALLY! Be inspired...

Make a list of different revision ideas and beside each one write **THINGS I HAVE** tried, **THINGS I WILL** try and **THINGS I MIGHT** try. Don't be scared of trying something new.

And remember – "FAIL TO PREPARE AND PREPARE TO FAIL!"

Intermediate 2 Physics

The course

The Intermediate 2 Physics course is designed to provide opportunities to develop knowledge and understanding of the concepts of physics and the ability to solve problems and to carry out experimental and investigative work. You are expected to have already reached the standards of Intermediate 1 Physics **and** Intermediate 1 Mathematics.

The course aims to enable you to:

- have an increased knowledge and understanding of facts and ideas, of techniques and of the applications of physics in society
- apply your knowledge and understanding in a wide variety of theoretical and practical problem solving contexts
- carry out experimental and investigative work in physics and analyse the information obtained.

The exam

Remember to bring the following to the exam:

- A calculator – treat yourself to new batteries!
- A pen, pencil, ruler and rubber. A ruler is essential for drawing light ray diagrams, vector diagrams and graphs, where marks can be lost because the lines are not straight.

Avoid using correction fluid – it is time consuming waiting for it to dry and candidates often forget to make the necessary correction after it has dried.

General advice

Read the questions thoroughly, note the information carefully and select the appropriate information. Your answers must also be clear and legible. Marks can be lost if the examiner cannot decipher what has been written. Avoid careless and minimal responses in the 'describe and explain' questions.

Make sure you know and understand the appropriate definitions and concepts as given in the Intermediate 2 Physics content statements which are available from the SQA website at www.sqa.org.uk. Definitions which are commonly required include ionisation, the activity of a radioactive substance, half-life, a.c. and d.c., critical angle and refraction.

Practise drawing electrical circuits, making sure you can place a *voltmeter in parallel* with a component to measure the *voltage across* it and an *ammeter in series* in the circuit to measure the *current through* it. Be able to draw and identify the correct symbols for the electrical components as detailed in the content statements e.g.

ammeter, voltmeter, battery, resistor etc.

When only one answer is required, do not be tempted to put more than one. Wrong answers cancel out correct ones!

Formulae and calculations

Become familiar with using the SQA Data sheet provided in the exams. This sheet is also available at the SQA website. Here are some useful tips.

1. Learn the formulae – even though they are provided in the data booklet. Make sure you understand what each one means and how and when they should be applied.

2. Symbols – use the symbols as stated in the formulae and **do not** substitute symbols such as 'C' for current instead of 'I'. Learn the meaning of the symbols. Note: the symbol for seconds is **s**, not **secs**.

3. Units – learn the units associated with each of the quantities. Remember to include units in the final answers, and check that they are the correct units. Weight is often answered in kg instead of N (newtons).

4. Gain practice using all of the prefixes listed in the content statements (μ, m, k, M, G) and be able to enter them into your calculator correctly. Also, do not attempt any unnecessary conversions, e.g. kilograms into grams. However, remember that km must be converted into metres and minutes into seconds.

5. Practise using scientific notation in your calculator e.g. 3×10^8. Do not be tempted to write out 300 000 000 as it is easy to miscount the number of noughts.

6. Significant figures – remember the general rule is that you are allowed between 1 less and 2 more than the smallest number in the data. Do not quote everything to two decimal places. A number such as 1063.54 has only two decimal places but it is six significant figures. Never use the recurring sign (i.e. a dot above the last number) – this is telling the examiner that you know the answer is accurate to an infinite number of figures!

7. Formulae displayed in triangles should be avoided. They will not get any marks if the proper formula is omitted.

8. A standard '2 mark answer' requires a formula (1/2 mark), correct substitution (1/2 mark) and a

numerical answer with the correct unit (1 mark). Naturally, you will achieve full marks by supplying the correct answer but you are at risk of losing a lot of marks if the full solution is not supplied and an arithmetic error has occurred.

9 Gain practice in rearranging the data in an equation to allow the unknown quantity to be calculated.

10 When an answer is worth 3 marks, there is usually an additional piece of information required on top of the standard 2 mark calculation.

11 Use common sense: if you calculate the speed of a car as 1×10^6 m/s or the mass of a person as 800 kg then you have made a mistake.

The Data Sheet

This is published at the beginning of the Intermediate 2 Paper.

1 Become familiar with the data and identify the most common data needed i.e. the speeds of light and sound in air; the gravitational field strength on Earth; the specific heat capacity of water etc.

2 Do not confuse *fusion* (liquid to solid) with *vaporisation* (liquid to gas).

3 Do not confuse the speed of sound, 340 m/s with the speed of light, 3×10^8 m/s. Note that this is also the speed of all the waves in the electromagnetic spectrum (radiowaves, microwaves, infrared radiation etc.)!

Graphs

1 When drawing a graph, make sure you use the majority of the grid space available. Label the axes carefully with both the name of the variable and the units.

2 Always think about the features of the graph: what do the axis intercepts, the gradient of the line and the area underneath the line mean – what is their physical significance?

3 When answering graph questions, be specific about what you are doing – state 'area under graph' or 'gradient of line' rather than just showing the calculations themselves. For example, the distance travelled by an object can be calculated from the area under a speed-time graph whilst the acceleration can be determined by calculating the gradient.

4 Read the axes carefully – is the scale in newtons (N) or kilonewtons (kN), seconds (s) or milliseconds (ms)?

Good luck!

Remember that the rewards for passing Intermediate 2 Physics are well worth it! Your pass will help you get the future you want for yourself. In the exam, be confident in your own ability. If you are not sure how to answer a question, trust your instincts and just give it a go anyway. Keep calm and don't panic! GOOD LUCK!

INTERMEDIATE 2

2009

[BLANK PAGE]

X069/201

NATIONAL QUALIFICATIONS 2009	TUESDAY, 26 MAY 1.00 PM – 3.00 PM	PHYSICS INTERMEDIATE 2

Read Carefully

Reference may be made to the Physics Data Booklet

1   All questions should be attempted.

Section A (questions 1 to 20)

2   Check that the answer sheet is for Physics Intermediate 2 (Section A).

3   For this section of the examination you must use an **HB pencil** and, where necessary, an eraser.

4   Check that the answer sheet you have been given has **your name**, **date of birth**, **SCN** (Scottish Candidate Number) and **Centre Name** printed on it.

    Do not change any of these details.

5   If any of this information is wrong, tell the Invigilator immediately.

6   If this information is correct, **print** your name and seat number in the boxes provided.

7   There is **only one correct** answer to each question.

8   Any rough working should be done on the question paper or the rough working sheet, **not** on your answer sheet.

9   At the end of the exam, put the **answer sheet for Section A inside the front cover of your answer book**.

10   Instructions as to how to record your answers to questions 1–20 are given on page three.

Section B (questions 21 to 29)

11   Answer the questions numbered 21 to 29 in the answer book provided.

12   **All answers must be written clearly and legibly in ink**.

13   Fill in the details on the front of the answer book.

14   Enter the question number clearly in the margin of the answer book beside each of your answers to questions 21 to 29.

15   Care should be taken to give an appropriate number of significant figures in the final answers to calculations.

DATA SHEET

Speed of light in materials

Material	Speed in m/s
Air	3.0×10^8
Carbon dioxide	3.0×10^8
Diamond	1.2×10^8
Glass	2.0×10^8
Glycerol	2.1×10^8
Water	2.3×10^8

Speed of sound in materials

Material	Speed in m/s
Aluminium	5200
Air	340
Bone	4100
Carbon dioxide	270
Glycerol	1900
Muscle	1600
Steel	5200
Tissue	1500
Water	1500

Gravitational field strengths

	Gravitational field strength on the surface in N/kg
Earth	10
Jupiter	26
Mars	4
Mercury	4
Moon	1.6
Neptune	12
Saturn	11
Sun	270
Venus	9

Specific heat capacity of materials

Material	Specific heat capacity in J/kg °C
Alcohol	2350
Aluminium	902
Copper	386
Glass	500
Ice	2100
Iron	480
Lead	128
Oil	2130
Water	4180

Specific latent heat of fusion of materials

Material	Specific latent heat of fusion in J/kg
Alcohol	0.99×10^5
Aluminium	3.95×10^5
Carbon dioxide	1.80×10^5
Copper	2.05×10^5
Iron	2.67×10^5
Lead	0.25×10^5
Water	3.34×10^5

Melting and boiling points of materials

Material	Melting point in °C	Boiling point in °C
Alcohol	−98	65
Aluminium	660	2470
Copper	1077	2567
Glycerol	18	290
Lead	328	1737
Iron	1537	2747

Specific latent heat of vaporisation of materials

Material	Specific latent heat of vaporisation in J/kg
Alcohol	11.2×10^5
Carbon dioxide	3.77×10^5
Glycerol	8.30×10^5
Turpentine	2.90×10^5
Water	22.6×10^5

Radiation weighting factors

Type of radiation	Radiation weighting factor
alpha	20
beta	1
fast neutrons	10
gamma	1
slow neutrons	3

SECTION A

For questions 1 to 20 in this section of the paper the answer to each question is either A, B, C, D or E. Decide what your answer is, then, using your pencil, put a horizontal line in the space provided—see the example below.

EXAMPLE

The energy unit measured by the electricity meter in your home is the

    A   kilowatt-hour

    B   ampere

    C   watt

    D   coulomb

    E   volt.

The correct answer is **A**—kilowatt-hour. The answer **A** has been clearly marked in **pencil** with a horizontal line (see below).

Changing an answer

If you decide to change your answer, carefully erase your first answer and, using your pencil, fill in the answer you want. The answer below has been changed to **E**.

[Turn over

SECTION A

Answer questions 1–20 on the answer sheet.

1. Which of the following quantities requires both magnitude and direction?

   A   Mass

   B   Distance

   C   Momentum

   D   Speed

   E   Time

2. A cross country runner travels 2·1 km North then 1·5 km East. The total time taken is 20 minutes.

   The average speed of the runner is

   A   0·18 m/s

   B   2·2 m/s

   C   3·0 m/s

   D   130 m/s

   E   180 m/s.

3. The graph shows how the velocity of an object varies with time.

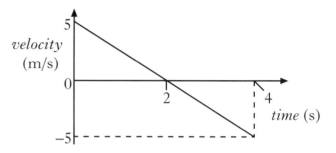

   Which row in the table shows the displacement after 4 s and the acceleration of the object during the first 4 s?

	Displacement (m)	Acceleration (m/s^2)
A	10	−10
B	10	2·5
C	0	2·5
D	0	−10
E	0	−2·5

4 A ball is thrown horizontally from a cliff as shown.

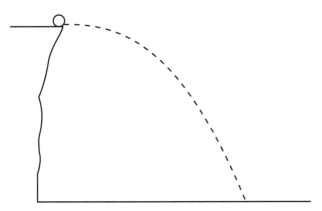

The effect of air resistance is negligible.

A student makes the following statements about the ball.

   I The vertical speed of the ball increases as it falls.

   II The vertical acceleration of the ball increases as it falls.

   III The vertical force on the ball increases as it falls.

Which of the statements is/are correct?

A I only

B II only

C I and II only

D II and III only

E I, II and III

5. Which block has the largest resultant force acting on it?

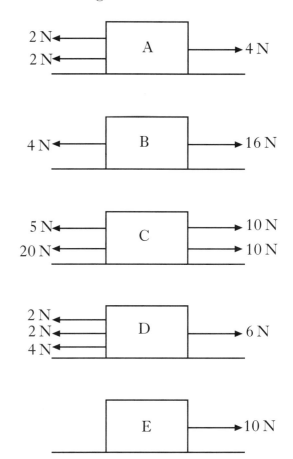

[Turn over

6. An arrow is fired from a bow as shown.

An archer pulls the string back a distance of 0·50 m. The string exerts an average force of 300 N on the arrow as it is fired. The mass of the arrow is 0·15 kg.

The maximum kinetic energy gained by the arrow is

A   23 J

B   150 J

C   600 J

D   2000 J

E   6750 J.

7. A solid substance is placed in an insulated container and is heated at a constant rate. The graph shows how the temperature of the substance changes with time.

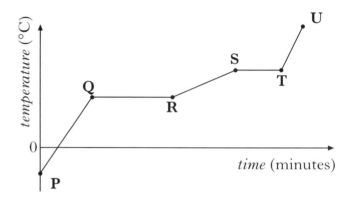

During the time interval QR, which of the following statements is/are correct?

   I   There is a change in the state of the substance.

  II   The substance changes state from a liquid to a gas.

III   Heat is absorbed by the substance.

A   I only

B   III only

C   I and II only

D   I and III only

E   I, II and III

8. A student writes the following statements about electrical conductors.

   I Only protons are free to move.

   II Only electrons are free to move.

   III Only negative charges are free to move.

Which of the statements is/are correct?

A   I only

B   II only

C   III only

D   I and II only

E   II and III only

9. A charge of 15 C passes through a resistor in 12 s. The potential difference across the resistor is 6 V.

The power developed by the resistor is

A   4·8 W

B   7·5 W

C   9·4 W

D   30 W

E   1080 W.

10. A circuit is set up as shown.

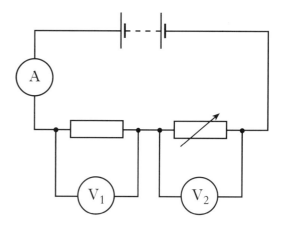

The resistance of the variable resistor is increased.

Which row in the table shows the effect on the readings on the ammeter and voltmeters?

	Reading on ammeter	Reading on voltmeter V_1	Reading on voltmeter V_2
A	decreases	decreases	decreases
B	increases	unchanged	increases
C	decreases	increases	decreases
D	increases	unchanged	decreases
E	decreases	decreases	increases

[Turn over

11. A circuit is set up as shown.

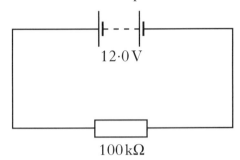

The power supplied to the resistor is

A 1.20×10^{-4} W

B 1.44×10^{-3} W

C 1.44 W

D 694 W

E 1.20×10^6 W.

12. Which of the following devices transforms light energy into electrical energy?

A   LED

B   Thermocouple

C   Microphone

D   Solar cell

E   Transistor

13. Which of the following is the correct symbol for an n-channel enhancement MOSFET?

A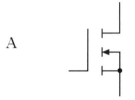

B

C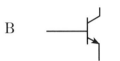

D

E

14. Which of the following is an example of a longitudinal wave?

    A   Light wave

    B   Infra-red wave

    C   Radio wave

    D   Sound wave

    E   Water wave

15. The diagram shows a list of the members of the electromagnetic spectrum in order of increasing wavelength.

gamma rays	P	ultraviolet	Q	infrared	R	TV and Radio

Which row in the table shows the radiation represented by the letters **P**, **Q** and **R**?

	P	**Q**	**R**
A	microwaves	visible light	x-rays
B	visible light	microwaves	x-rays
C	x-rays	visible light	microwaves
D	visible light	x-rays	microwaves
E	x-rays	microwaves	visible light

16. The diagram shows what happens to a ray of light when it strikes a glass block.

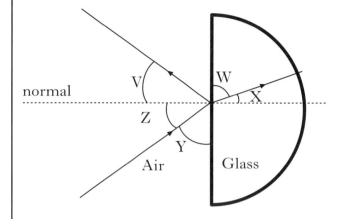

Which row in the table identifies the angle of incidence and the angle of refraction?

	Angle of Incidence	Angle of Refraction
A	V	W
B	Y	W
C	Y	X
D	Z	W
E	Z	X

[Turn over

17. The diagram below shows a simple model of an atom.

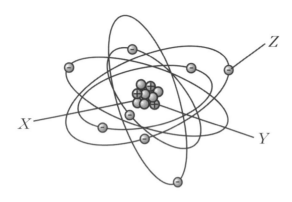

Which row in the table identifies particles *X*, *Y* and *Z*?

	X	*Y*	*Z*
A	electron	proton	neutron
B	proton	neutron	electron
C	neutron	electron	proton
D	electron	neutron	proton
E	neutron	proton	electron

18. A student makes the following statements about ionising radiations.

    I Ionisation occurs when an atom loses an electron.

    II Gamma radiation produces greater ionisation (density) than alpha particles.

    III An alpha particle consists of 2 protons, 2 neutrons and 2 electrons.

    Which of the statements is/are correct?

    A I only

    B II only

    C I and II only

    D II and III only

    E I, II and III

19. A sample of tissue has a mass of 0·05 kg.

    The tissue is exposed to radiation and absorbs 0·1 J of energy in 2 minutes.

    The absorbed dose is

    A 0·005 Gy

    B 0·1 Gy

    C 0·5 Gy

    D 2 Gy

    E 6 Gy.

20. During fission, a neutron splits a uranium nucleus into two nuclei, X and Y, as shown below.

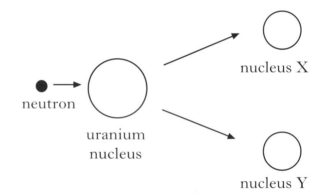

    For a chain reaction to occur which of the following **must** also be released?

    A Protons

    B Electrons

    C Neutrons

    D Alpha particles

    E Gamma radiation

Candidates are reminded that the answer sheet for Section A MUST be placed INSIDE the front cover of the answer book.

SECTION B

Marks

Write your answers to questions 21–29 in the answer book.

All answers must be written clearly and legibly in ink.

21. A ski lift with a gondola of mass 2000 kg travels to a height of 540 m from the base station to a station at the top of the mountain.

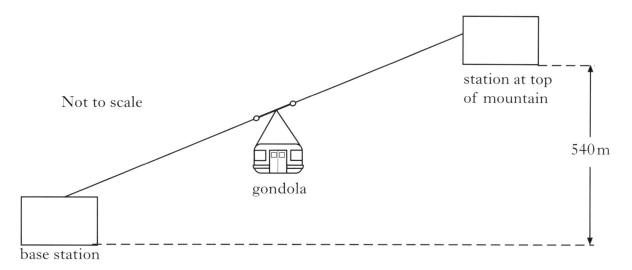

(a) Calculate the gain in gravitational potential energy of the gondola.

2

(b) During the journey, the kinetic energy of the gondola is 64 000 J.

Calculate the speed of the gondola.

2

(c) The ski lift requires a motor which operates at 380 V to take the gondola up the mountain. The maximum power produced is 45·6 kW.

(i) Calculate the maximum current in the motor.

2

(ii) Calculate the electrical energy used by the motor when it has been operating at its maximum power for a total time of 1 hour.

2

(8)

[Turn over

Marks

22. A child sledges down a hill.

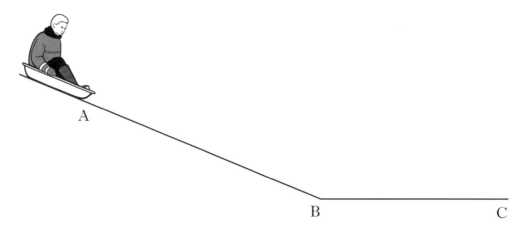

The sledge and child are released from rest at point A. They reach a speed of 3 m/s at point B.

(a) The sledge and child take 5 s to reach point B.

Calculate the acceleration.   **2**

(b) The sledge and child have a combined mass of 40 kg.

Calculate the unbalanced force acting on them.   **2**

(c) After the sledge and child pass point B, they slow down, coming to a halt at point C.

Explain this motion in terms of forces.   **2**

   (6)

Marks

23. The following apparatus is used to determine the speed of a pellet as it leaves an air rifle. The air rifle fires a pellet into the plasticine, causing the vehicle to move.

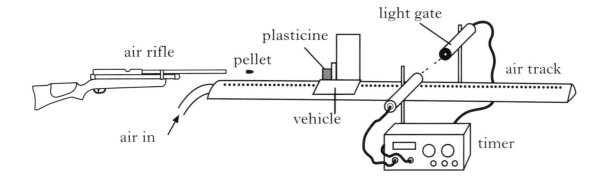

(a) Describe how the apparatus is used to determine the speed of the vehicle.

Your description must include:

- the measurements made
- any necessary calculations.   2

(b) The speed of the vehicle is calculated as 0·35 m/s after impact.

The mass of the pellet is $5·0 \times 10^{-4}$ kg. The mass of the vehicle and plasticine before impact is 0·30 kg.

   (i) Show that the momentum of the pellet **before** impact with the plasticine is 0·105 kg m/s.   1

   (ii) Hence, calculate the velocity of the pellet **before** impact with the plasticine.   1

(c) At a firing range a pellet is fired horizontally at a target 40 m away. It takes 0·20 s to reach the target.

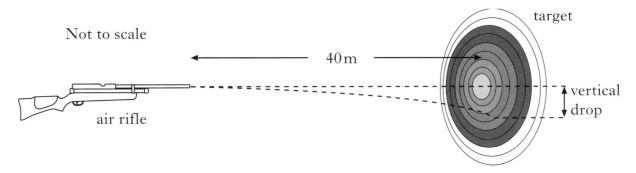

   (i) Calculate the **vertical** velocity of the pellet on reaching the target.   2

   (ii) Calculate the vertical drop.   2

    (8)

[Turn over

Marks

24. A fridge/freezer has water and ice dispensers as shown.

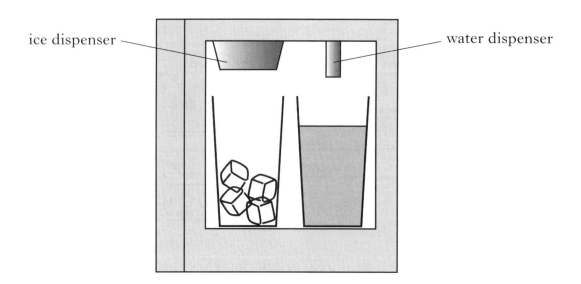

ice dispenser   water dispenser

(*a*) Water of mass 0·1 kg flows into the freezer at 15 °C and is cooled to 0 °C. Calculate the energy removed when the water cools.   **2**

(*b*) Calculate how much energy is released when 0·1 kg of water at 0 °C changes to 0·1 kg of ice at 0 °C.   **2**

(*c*) The fridge/freezer system removes heat energy at a rate of 125 J/s.

   (i) Calculate the minimum time taken to produce 0·1 kg of ice from 0·1 kg of water at 15 °C.   **3**

   (ii) Explain why the actual time taken to make the ice will be longer than the value calculated in part (i).   **2**

(9)

Marks

25. A student sets up the following circuit to investigate the resistance of resistor R.

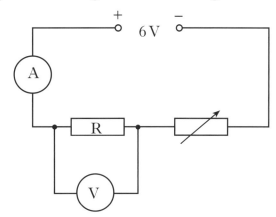

The variable resistor is adjusted and the voltmeter and ammeter readings are noted. The following graph is obtained from the experimental results.

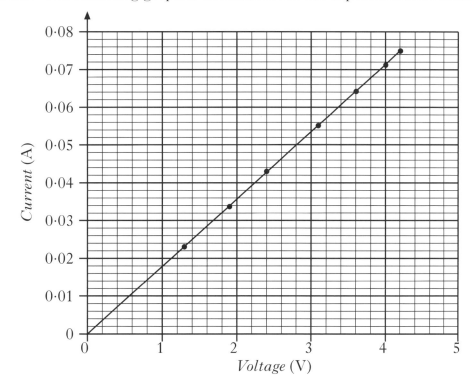

(a) (i) Calculate the value of the resistor R when the reading on the voltmeter is 4·2 V. 3

(ii) Using information from the graph, state whether the resistance of the resistor R, **increases**, **stays the same** or **decreases** as the voltage increases.

Justify your answer. 2

(b) The student is given a task to combine two resistors from a pack containing one each of 33 Ω, 56 Ω, 82 Ω, 150 Ω, 270 Ω, 390 Ω.

Show by calculation which **two** resistors should be used to give:

(i) the largest combined resistance; 2

(ii) the smallest combined resistance. 2

(9)

Marks

26. An MP3 player is charged from the mains supply of 230 V using a transformer, which has an output voltage of 5 V and an output current of 1 A.

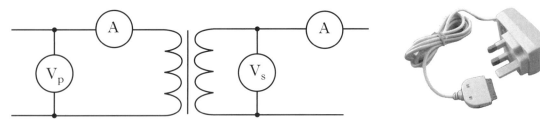

circuit diagram of transformer   MP3 Charger

(a) Calculate the current in the primary circuit.   2

(b) The MP3 player is then put on a docking station with external speakers.

docking station with speakers   MP3 player

    (i) Calculate the resistance of a 10 W speaker when the voltage across it is 9 V.   2

    (ii) Calculate the gain of the amplifier in the docking station when the input voltage is 1·5 V.   2

(c) The input power to the amplifier is 25 W. The output power is 20 W. Calculate the efficiency of the amplifier.   2

           (8)

Marks

27. A student is short sighted.

    (*a*)   (i) What does the term "short sighted" mean?   **1**

         (ii) What type of lens is required to correct this eye defect?   **1**

         (iii) The focal length of the lens needed to correct the student's short sight is 180 mm. Calculate the power of this lens.   **2**

    (*b*) In the eye, refraction of light occurs at both the cornea and the lens. Some eye defects can be corrected using a laser. Light from the laser is used to change the shape of the cornea.

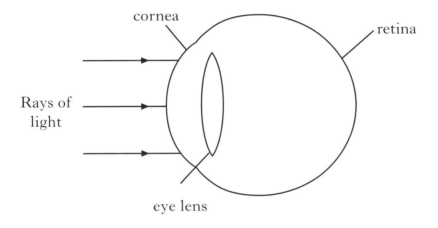

         (i) State what is meant by refraction of light.   **1**

         (ii) The laser emits light of wavelength 7×10^{-7} m.

            Calculate the frequency of the light.   **2**

    (*c*) Lasers can be used in optical fibres for medical purposes.

         (i) Copy and complete the path of the laser light along the optical fibre.   **2**

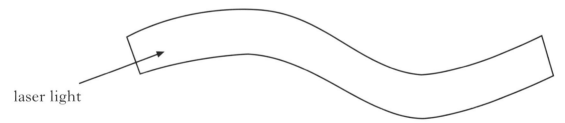

laser light

         (ii) Name the effect when the laser light hits the inside surface of the fibre.   **1**

                                                      (10)

[Turn over

Marks

28. Parking sensors are fitted to the rear bumper of some cars. A buzzer emits audible beeps, which become more frequent as the car moves closer to an object.

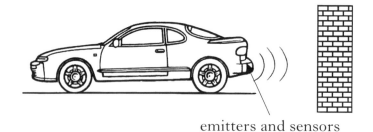

emitters and sensors

Ultrasonic pulses are emitted from the rear of the car. Objects behind the car reflect the pulses, which are detected by sensors. Ultrasonic pulses travel at the speed of sound.

(a) The time between these pulses being sent and received is 2×10^{-3} s.

Calculate the distance between the object and the rear of the car.   **3**

(b) At a certain distance, the buzzer beeps every 0·125 s.

Calculate the frequency of the beeps.   **2**

(c) The sensor operates at a voltage of 12 V and has a current range of 20–200 mA.

Calculate the maximum power rating of the sensor.   **3**

(d) An LED system can be added so that it flashes at the same frequency as the beeps from the buzzer. The LED circuit is shown below.

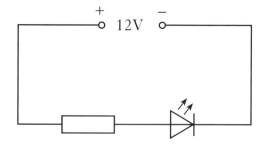

(i) A resistor is connected in series with the LED.

State the purpose of the resistor.   **1**

(ii) When lit, the LED has a voltage of 3·5 V across it and a current of 200 mA.

Calculate the value of the resistor.   **3**

    (12)

Marks

29. A radioactivity kit includes three radioactive sources each made up as shown.

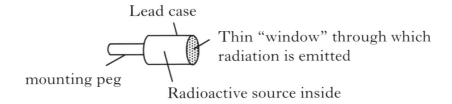

Lead case

Thin "window" through which radiation is emitted

mounting peg

Radioactive source inside

Information about these sources is given in the table below.

	Radiation Emitted	*Radioactive Element*
Source 1	Alpha	Americium 241
Source 2	Beta	Strontium 90
Source 3	Gamma + Beta	Cobalt 60

(*a*) (i) Describe an experiment to show which is the alpha emitting source.

Your description must include:

- equipment used

- measurements taken

- an explanation of the results. **3**

(ii) The radioactive material in Source 3 emits both beta and gamma radiations. Describe how the window of the casing could be modified so that the beta radiation is stopped. **1**

(*b*) Strontium 90 has a half life of 28 years. Calculate how many years it takes for the activity to decrease to 1/16th of its original value. **2**

(*c*) (i) A technician working with Source 1 receives an absorbed dose of $20\,\mu$Gy of alpha particles. Calculate the total equivalent dose received by the technician. **2**

(ii) Describe two ways in which the technician could reduce his absorbed dose. **2**

(10)

[END OF QUESTION PAPER]

[BLANK PAGE]

INTERMEDIATE 2

2010

[BLANK PAGE]

X069/201

NATIONAL QUALIFICATIONS 2010	FRIDAY, 28 MAY 1.00 PM – 3.00 PM	PHYSICS INTERMEDIATE 2

Read Carefully

Reference may be made to the Physics Data Booklet

1 All questions should be attempted.

Section A (questions 1 to 20)

2 Check that the answer sheet is for Physics Intermediate 2 (Section A).

3 For this section of the examination you must use an **HB pencil** and, where necessary, an eraser.

4 Check that the answer sheet you have been given has **your name**, **date of birth**, **SCN** (Scottish Candidate Number) and **Centre Name** printed on it.

   Do not change any of these details.

5 If any of this information is wrong, tell the Invigilator immediately.

6 If this information is correct, **print** your name and seat number in the boxes provided.

7 There is **only one correct** answer to each question.

8 Any rough working should be done on the question paper or the rough working sheet, **not** on your answer sheet.

9 At the end of the exam, put the **answer sheet for Section A inside the front cover of your answer book**.

10 Instructions as to how to record your answers to questions 1–20 are given on page three.

Section B (questions 21 to 30)

11 Answer the questions numbered 21 to 30 in the answer book provided.

12 **All answers must be written clearly and legibly in ink**.

13 Fill in the details on the front of the answer book.

14 Enter the question number clearly in the margin of the answer book beside each of your answers to questions 21 to 30.

15 Care should be taken to give an appropriate number of significant figures in the final answers to calculations.

DATA SHEET

Speed of light in materials

Material	Speed in m/s
Air	3.0×10^8
Carbon dioxide	3.0×10^8
Diamond	1.2×10^8
Glass	2.0×10^8
Glycerol	2.1×10^8
Water	2.3×10^8

Speed of sound in materials

Material	Speed in m/s
Aluminium	5200
Air	340
Bone	4100
Carbon dioxide	270
Glycerol	1900
Muscle	1600
Steel	5200
Tissue	1500
Water	1500

Gravitational field strengths

	Gravitational field strength on the surface in N/kg
Earth	10
Jupiter	26
Mars	4
Mercury	4
Moon	1·6
Neptune	12
Saturn	11
Sun	270
Venus	9

Specific heat capacity of materials

Material	Specific heat capacity in J/kg °C
Alcohol	2350
Aluminium	902
Copper	386
Glass	500
Ice	2100
Iron	480
Lead	128
Oil	2130
Water	4180

Specific latent heat of fusion of materials

Material	Specific latent heat of fusion in J/kg
Alcohol	0.99×10^5
Aluminium	3.95×10^5
Carbon dioxide	1.80×10^5
Copper	2.05×10^5
Iron	2.67×10^5
Lead	0.25×10^5
Water	3.34×10^5

Melting and boiling points of materials

Material	Melting point in °C	Boiling point in °C
Alcohol	−98	65
Aluminium	660	2470
Copper	1077	2567
Glycerol	18	290
Lead	328	1737
Iron	1537	2747

Specific latent heat of vaporisation of materials

Material	Specific latent heat of vaporisation in J/kg
Alcohol	11.2×10^5
Carbon dioxide	3.77×10^5
Glycerol	8.30×10^5
Turpentine	2.90×10^5
Water	22.6×10^5

Radiation weighting factors

Type of radiation	Radiation weighting factor
alpha	20
beta	1
fast neutrons	10
gamma	1
slow neutrons	3

SECTION A

For questions 1 to 20 in this section of the paper the answer to each question is either A, B, C, D or E. Decide what your answer is, then, using your pencil, put a horizontal line in the space provided—see the example below.

EXAMPLE

The energy unit measured by the electricity meter in your home is the

    A kilowatt-hour

    B ampere

    C watt

    D coulomb

    E volt.

The correct answer is **A**—kilowatt-hour. The answer **A** has been clearly marked in **pencil** with a horizontal line (see below).

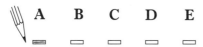

Changing an answer

If you decide to change your answer, carefully erase your first answer and, using your pencil, fill in the answer you want. The answer below has been changed to **E**.

A   B   C   D   E

[Turn over

SECTION A

Answer questions 1–20 on the answer sheet.

1. Which of the following is a scalar quantity?

    A   Force

    B   Acceleration

    C   Momentum

    D   Velocity

    E   Energy

2. A student investigates the speed of a trolley as it moves down a slope.

    The apparatus is set up as shown.

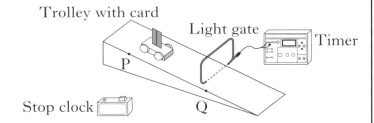

    The following measurements are recorded.

    distance from P to Q = 1·0 m
    length of card on trolley = 0·04 m
    time taken for trolley to travel from P to Q = 2·5 s
    time taken for card to pass through light gate = 0·05 s

    The speed at Q is

    A   0·002 m/s

    B   0·016 m/s

    C   0·40 m/s

    D   0·80 m/s

    E   20 m/s.

3. Two forces, each of 7 N, act on an object O.

    The forces act as shown.

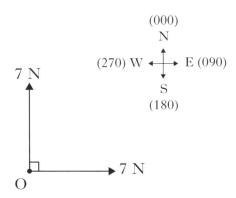

    The resultant of these two forces is

    A   7 N at a bearing of 135

    B   9·9 N at a bearing of 045

    C   9·9 N at a bearing of 135

    D   14 N at a bearing of 045

    E   14 N at a bearing of 135.

4 A package is released from a helicopter flying horizontally at a constant speed of 40 m/s.

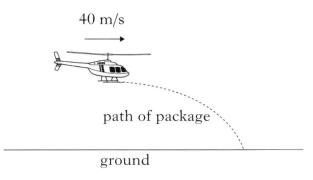

The package takes 3·0 s to reach the ground.

The effects of air resistance can be ignored.

Which row in the table shows the horizontal speed and vertical speed of the package just before it hits the ground?

	Horizontal speed (m/s)	Vertical speed (m/s)
A	0	30
B	30	30
C	30	40
D	40	30
E	40	40

5. 100 g of a solid is heated by a 50 W heater. The graph of temperature of the substance against time is shown.

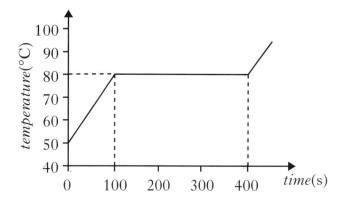

The specific latent heat of fusion of the substance is

A $1\cdot3 \times 10^3$ J/kg

B $1\cdot5 \times 10^3$ J/kg

C $3\cdot0 \times 10^3$ J/kg

D $1\cdot5 \times 10^5$ J/kg

E $1\cdot9 \times 10^5$ J/kg.

[Turn over

6. A crate of mass 200 kg is pushed a distance of 20 m across a level floor.

The crate is pushed with a force of 150 N.

The force of friction acting on the crate is 50 N.

The work done in pushing the crate across the floor is

A   1000 J

B   2000 J

C   3000 J

D   4000 J

E   20 000 J.

7. A student makes the following statements about electrical circuits.

I The sum of the potential differences across components connected in series is equal to the supply voltage.

II The sum of the currents in parallel branches is equal to the current drawn from the supply.

III The potential difference across components connected in parallel is the same for each component.

Which of the statements is/are correct?

A   I only

B   III only

C   I and II only

D   II and III only

E   I, II and III

8. Three resistors are connected as shown

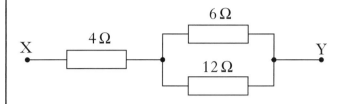

The total resistance between X and Y is

A   2 Ω

B   4 Ω

C   8 Ω

D   13 Ω

E   22 Ω.

9. The resistance of a wire is 6 Ω.

The current in the wire is 2 A.

The power developed in the wire is

A   3 W

B   12 W

C   18 W

D   24 W

E   72 W.

10. The voltage of the mains supply in the UK is 230 V a.c.

Which row in the table shows the peak voltage and frequency of the mains supply in the UK?

	peak voltage (V)	frequency (Hz)
A	175	50
B	175	60
C	230	50
D	325	50
E	325	60

11. The diagram shows a model bicycle dynamo.

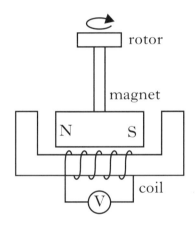

When the rotor is turned the magnet rotates, inducing a voltage in the coil. The induced voltage can be decreased by

A   increasing the number of turns on the coil

B   decreasing the number of turns on the coil

C   using a stronger magnet

D   turning the rotor faster

E   reversing the direction of rotation of the magnet.

12. The graph below shows how the input voltage V_1 to an amplifier varies with time t.

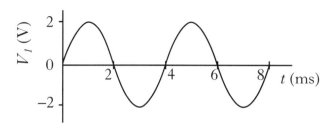

The amplifier has a voltage gain of 10.

Which graph shows how the output voltage V_0 of the amplifier varies with time t?

A

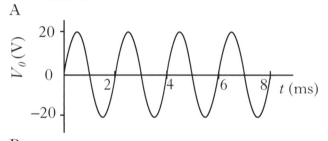

B

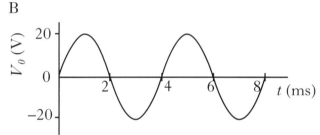

C

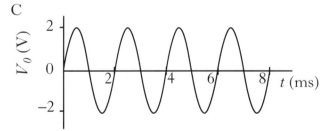

D

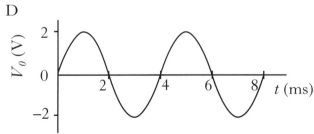

E

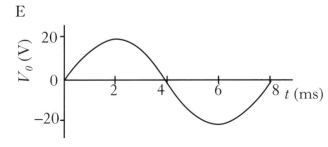

13. The diagram gives information about a wave.

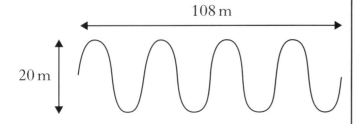

The time taken for the waves to travel 108 m is 0·5 s.

A student makes the following statements about the waves.

  I   The wavelength of the waves is 27 m.

  II   The amplitude of the waves is 20 m.

  III   The frequency of the waves is 8 Hz.

Which of the statements is/are correct?

A   I only

B   II only

C   I and III only

D   II and III only

E   I, II and III

14. The diagram shows members of the electromagnetic spectrum in order of increasing wavelength.

Gamma rays	P	Ultraviolet radiation	Q	Infrared radiation	R	TV & radio waves

———— increasing wavelength ————→

Which row in the table identifies the radiations represented by the letters P, Q and R?

	P	Q	R
A	X-rays	visible light	microwaves
B	X-rays	microwaves	visible light
C	microwaves	visible light	X-rays
D	visible light	microwaves	X-rays
E	visible light	X-rays	microwaves

15. An object is placed in front of a converging lens as shown.

The position of the image formed by the lens is also shown.

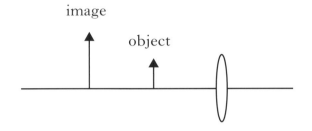

The focal length of the lens is 100 mm.

The distance between the lens and the object is

A　50 mm

B　100 mm

C　150 mm

D　200 mm

E　250 mm.

16. A converging lens has a focal length of 50 mm.

The power of the lens is

A　+0·02 D

B　+0·2 D

C　−0·2 D

D　+20 D

E　−20 D.

17. A student makes the following statements about a carbon atom.

　I The atom is made up only of protons and neutrons.

　II The nucleus of the atom contains protons, neutrons and electrons.

　III The nucleus of the atom contains only protons and neutrons.

Which of the statements is/are correct?

A　I only

B　II only

C　III only

D　I and II only

E　I and III only

18. Human tissue can be damaged by exposure to radiation.

On which of the following factors does the risk of biological harm depend?

　I The absorbed dose.

　II The type of radiation.

　III The body organs or tissue exposed.

A　I only

B　I and II only

C　II only

D　II and III only

E　I, II and III

[Turn over

19. Information about a radioactive source is given in Table 1.

Table 1

Activity	Energy absorbed per kilogram of tissue	Radiation weighting factor
500 MBq	0·2 μJ	10

Which row in Table 2 gives the correct information for the radioactive source?

Table 2

	Absorbed Dose	Equivalent Dose
A	0·2 μGy	2 μSv
B	500 MGy	10 Sv
C	10 Gy	0·2 μSv
D	20 μGy	50 MSv
E	2 μGy	0·2 μSv

20. In a nuclear reactor a chain reaction releases energy from nuclei.

Which of the following statements describes the beginning of a chain reaction?

A An electron splits a nucleus releasing more electrons.

B An electron splits a nucleus releasing protons.

C A proton splits a nucleus releasing more protons.

D A neutron splits a nucleus releasing electrons.

E A neutron splits a nucleus releasing more neutrons.

Candidates are reminded that the answer sheet for Section A MUST be placed INSIDE the front cover of the answer book.

SECTION B

Write your answers to questions 21–30 in the answer book.

All answers must be written clearly and legibly in ink.

21. A balloon of mass 400 kg rises vertically from the ground.

The graph shows how the vertical speed of the balloon changes during the first 100 s of its upward flight.

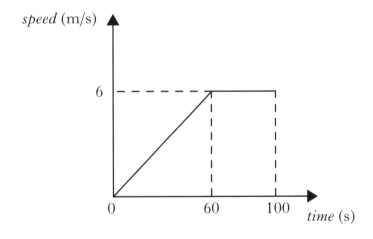

(a) Calculate the acceleration of the balloon during the first 60 s.   2

(b) Calculate the distance travelled by the balloon in 100 s.   2

(c) Calculate the average speed of the balloon during the first 100 s.   2

(d) Calculate the weight of the balloon.   2

(e) Calculate the total upward force acting on the balloon during the first 60 s of its flight.   3

(11)

Marks

22. Inside a storm cloud water droplets move around and collide with each other.

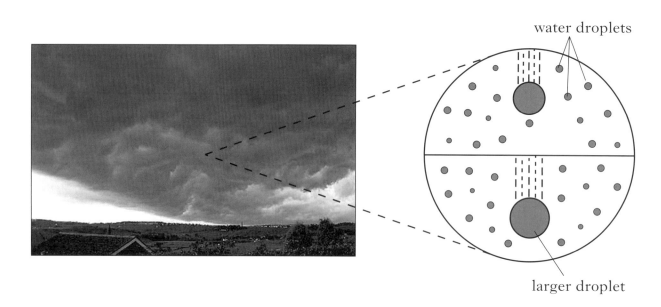

water droplets

larger droplet

(a) A water droplet of mass 2·0 g moving at a speed of 4·0 m/s collides with a stationary water droplet of mass 1·2 g. The two droplets join together to form a larger droplet.

Calculate the speed of this larger droplet after the collision.   2

(b) Another water droplet within the cloud is falling with a constant speed. Draw a diagram showing the forces acting on this droplet.

Name these forces and show their directions.   2

(c) The motion of water droplets in the cloud causes flashes of lightning. One lightning flash transfers 1650 C of charge in 0·15 s.

Calculate the electric current produced by this flash.   2

(d) Why does an observer, standing 3 km from a thunder cloud, see a lightning flash before he hears the thunder?   1

(7)

Marks

23. On the planet Mercury the surface temperature at night is −173 °C. The surface temperature during the day is 307 °C. A rock lying on the surface of the planet has a mass of 60 kg.

(a) The rock absorbs $2 \cdot 59 \times 10^7$ J of heat energy from the Sun during the day.

Calculate the specific heat capacity of the rock. 2

(b) Heat is released at a steady rate of 1440 J/s at night.

Calculate the time taken for the rock to release $2 \cdot 59 \times 10^7$ J of heat. 2

(c) Energy from these rocks could be used to heat a base on the surface of Mercury.

How many 60 kg rocks would be needed to supply a 288 kW heating system? 2

(d) Using information from the data sheet, would it be **easier**, **the same** or **more difficult** to lift rocks on Mercury compared to Earth?

You **must** explain your answer. 2

(8)

[Turn over

Marks

24. A student sets up the following circuit.

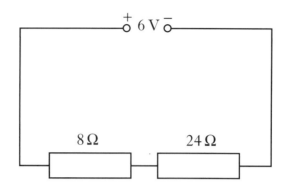

(a) Calculate the current in the $8\,\Omega$ resistor. **3**

(b) Calculate the voltage across the $8\,\Omega$ resistor. **2**

(c) The $24\,\Omega$ resistor is replaced by one of **greater** resistance. How will this affect the voltage across the $8\,\Omega$ resistor?

Explain your answer. **2**

(7)

Marks

25. In a lab experiment a technician builds a transformer and uses electrical meters to take a number of measurements, as shown in the diagram.

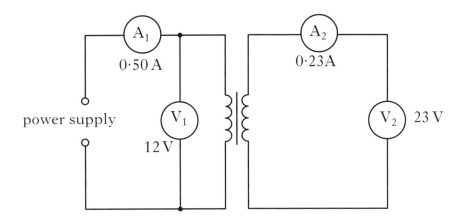

(a) The technician has a choice of an a.c. or a d.c. power supply. Which power supply should be used?

Explain your answer. 2

(b) Calculate the electrical power in the primary circuit of the transformer. 2

(c) Calculate the electrical power in the secondary circuit of the transformer. 1

(d) Calculate the percentage efficiency of the transformer. 2

(e) Another experiment uses a different transformer. It is 100% efficient. The primary coil has 1500 turns and the secondary coil contains 3000 turns.

Calculate the secondary voltage when the primary voltage is 12 V. 2

(9)

[Turn over

Marks

26. Water in a fish tank has to be maintained at a constant temperature. Part of the electronic circuit which controls the temperature is shown.

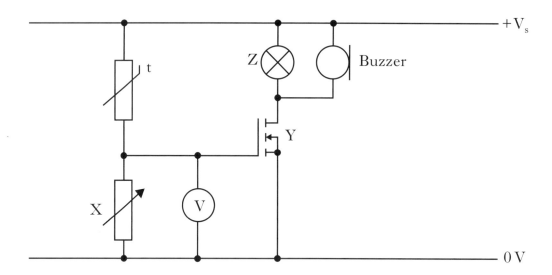

(a) Name components Y and Z.   **2**

(b) What happens to the resistance of the thermistor as the temperature increases?   **1**

(c) When the voltmeter reading reaches 1·8 V component Y switches on. Explain how the circuit operates when the temperature rises.   **2**

(d) Why is a variable resistor chosen for component X rather than a fixed value resistor?   **1**

(6)

Marks

27. At the kick-off in a football match, during the World Cup Finals, the referee blows his whistle. The whistle produces sound waves.

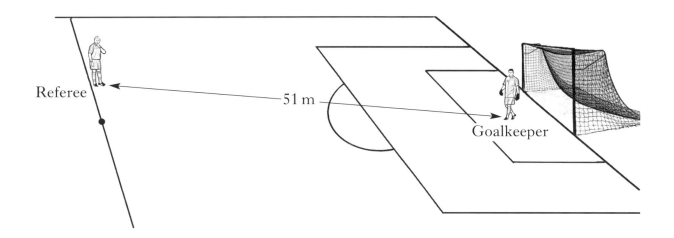

(a) Using information from the diagram and the data sheet, calculate the time taken for the sound waves to reach the goalkeeper.   2

(b)   (i) Are sound waves transverse or longitudinal waves?   1

     (ii) Describe **two** differences between transverse and longitudinal waves.   2

     (iii) What is transferred by waves?   1

(c)   (i) Floodlights in the stadium are switched on. Each lamp has a power rating of 2·40 kW. The operating voltage is 315 V.

      Calculate the resistance of a lamp.   2

     (ii) The floodlights consist of 20 lamps connected in parallel.

      State **two** reasons why the lamps are connected in parallel.   2

                                               (10)

[Turn over

Marks

28.   A satellite sends microwaves to a ground station on Earth.

(*a*) The microwaves have a wavelength of 60 mm.

(i) Calculate the frequency of the waves.   2

(ii) Determine the period of the waves.   2

(*b*) The satellite sends radio waves along with the microwaves to the ground station. Will the radio waves be received by the ground station **before**, **after** or **at the same time** as the microwaves?

Explain your answer.   2

(*c*) When the microwaves reach the ground station they are received by a curved reflector.

Explain why a curved reflector is used.

Your answer may include a diagram.   2

(8)

Marks

29. In 1908 Ernest Rutherford conducted a series of experiments involving alpha particles.

(a) State what is meant by an alpha particle. 1

(b) Alpha particles produce a greater ionisation density than beta particles or gamma rays. What is meant by the term *ionisation*? 1

(c) A radioactive source emits alpha particles and has a half-life of 2·5 hours. The source has an initial activity of 4·8 kBq.

Calculate the time taken for its activity to decrease to 300 Bq. 2

(d) Calculate the number of decays in the sample in two minutes, when the activity of the source is 1·2 kBq. 2

(e) Some sources emit alpha particles and are stored in lead cases despite the fact that alpha particles cannot penetrate paper. Suggest a possible reason for storing these sources using this method. 1

(7)

[Turn over for Question 30 on *Page twenty*

Marks

30. Many countries use nuclear reactors to produce energy. A diagram of the core of a nuclear reactor is shown.

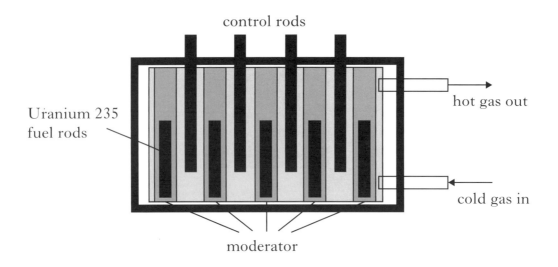

(a) State the purpose of:

   (i) the moderator;   **1**

   (ii) the control rods.   **1**

(b) One nuclear fission reaction produces $2 \cdot 9 \times 10^{-11}$ J of energy. The power output of the reactor is $1 \cdot 4$ GW. How many fission reactions are produced in one hour?   **3**

(c) State **one advantage** and **one disadvantage** of using nuclear power for the generation of electricity.   **2**

     (7)

[END OF QUESTION PAPER]

INTERMEDIATE 2

2011

[BLANK PAGE]

X069/201

NATIONAL QUALIFICATIONS 2011	MONDAY, 23 MAY 1.00 PM – 3.00 PM	PHYSICS INTERMEDIATE 2

Read Carefully

Reference may be made to the Physics Data Booklet

1   All questions should be attempted.

Section A (questions 1 to 20)

2   Check that the answer sheet is for Physics Intermediate 2 (Section A).

3   For this section of the examination you must use an **HB pencil** and, where necessary, an eraser.

4   Check that the answer sheet you have been given has **your name**, **date of birth**, **SCN** (Scottish Candidate Number) and **Centre Name** printed on it.

    Do not change any of these details.

5   If any of this information is wrong, tell the Invigilator immediately.

6   If this information is correct, **print** your name and seat number in the boxes provided.

7   There is **only one correct** answer to each question.

8   Any rough working should be done on the question paper or the rough working sheet, **not** on your answer sheet.

9   At the end of the exam, **put the answer sheet for Section A inside the front cover of your answer book**.

10   Instructions as to how to record your answers to questions 1–20 are given on page three.

Section B (questions 21 to 31)

11   Answer the questions numbered 21 to 31 in the answer book provided.

12   **All answers must be written clearly and legibly in ink**.

13   Fill in the details on the front of the answer book.

14   Enter the question number clearly in the margin of the answer book beside each of your answers to questions 21 to 31.

15   Care should be taken to give an appropriate number of significant figures in the final answers to calculations.

DATA SHEET

Speed of light in materials

Material	Speed in m/s
Air	3.0×10^8
Carbon dioxide	3.0×10^8
Diamond	1.2×10^8
Glass	2.0×10^8
Glycerol	2.1×10^8
Water	2.3×10^8

Speed of sound in materials

Material	Speed in m/s
Aluminium	5200
Air	340
Bone	4100
Carbon dioxide	270
Glycerol	1900
Muscle	1600
Steel	5200
Tissue	1500
Water	1500

Gravitational field strengths

	Gravitational field strength on the surface in N/kg
Earth	10
Jupiter	26
Mars	4
Mercury	4
Moon	1.6
Neptune	12
Saturn	11
Sun	270
Venus	9

Specific heat capacity of materials

Material	Specific heat capacity in J/kg °C
Alcohol	2350
Aluminium	902
Copper	386
Glass	500
Ice	2100
Iron	480
Lead	128
Oil	2130
Water	4180

Specific latent heat of fusion of materials

Material	Specific latent heat of fusion in J/kg
Alcohol	0.99×10^5
Aluminium	3.95×10^5
Carbon Dioxide	1.80×10^5
Copper	2.05×10^5
Iron	2.67×10^5
Lead	0.25×10^5
Water	3.34×10^5

Melting and boiling points of materials

Material	Melting point in °C	Boiling point in °C
Alcohol	−98	65
Aluminium	660	2470
Copper	1077	2567
Glycerol	18	290
Lead	328	1737
Iron	1537	2737

Specific latent heat of vaporisation of materials

Material	Specific latent heat of vaporisation in J/kg
Alcohol	11.2×10^5
Carbon Dioxide	3.77×10^5
Glycerol	8.30×10^5
Turpentine	2.90×10^5
Water	22.6×10^5

Radiation weighting factors

Type of radiation	Radiation weighting factor
alpha	20
beta	1
fast neutrons	10
gamma	1
slow neutrons	3

SECTION A

For questions 1 to 20 in this section of the paper the answer to each question is either A, B, C, D or E. Decide what your answer is, then, using your pencil, put a horizontal line in the space provided—see the example below.

EXAMPLE

The energy unit measured by the electricity meter in your home is the

    A kilowatt-hour

    B ampere

    C watt

    D coulomb

    E volt.

The correct answer is **A**—kilowatt-hour. The answer **A** has been clearly marked in **pencil** with a horizontal line (see below).

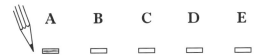

Changing an answer

If you decide to change your answer, carefully erase your first answer and, using your pencil, fill in the answer you want. The answer below has been changed to **E**.

    A   B   C   D   E

[Turn over

SECTION A

Answer questions 1–20 on the answer sheet.

1. During training an athlete sprints 30 m East and then 40 m West.

Which row shows the distance travelled and the displacement from the starting point?

	Distance travelled	Displacement
A	10 m	10 m East
B	10 m	10 m West
C	10 m	70 m East
D	70 m	10 m West
E	70 m	10 m East

2. The graph shows how the velocity of a ball changes with time.

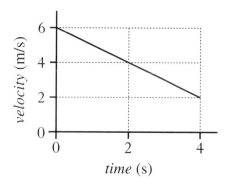

The acceleration of the ball is

A $-8\,\text{m/s}^2$

B $-1\,\text{m/s}^2$

C $1\,\text{m/s}^2$

D $8\,\text{m/s}^2$

E $24\,\text{m/s}^2$.

3. A ball of mass 2 kg moves along a horizontal surface at 4 m/s.

Which row shows the momentum and kinetic energy of the ball?

	Momentum (kg m/s)	Kinetic energy (J)
A	2	4
B	4	8
C	4	16
D	8	8
E	8	16

4. An engine applies a force of 2000 N to move a lorry at a constant speed.

The lorry travels 100 m in 16 s.

The power developed by the engine is

A $0{\cdot}8\,\text{W}$

B $12{\cdot}5\,\text{W}$

C $320\,\text{W}$

D $12\,500\,\text{W}$

E $3\,200\,000\,\text{W}$.

5. Which row in the table identifies the following circuit symbols?

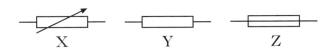

	Symbol X	Symbol Y	Symbol Z
A	fuse	resistor	variable resistor
B	fuse	variable resistor	resistor
C	resistor	variable resistor	fuse
D	variable resistor	fuse	resistor
E	variable resistor	resistor	fuse

6. Which graph shows how the potential difference V across a resistor varies with the current I in the resistor?

A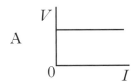

B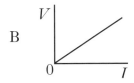

C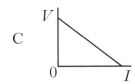

D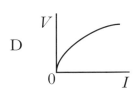

E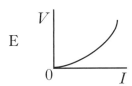

7. A circuit is set up as shown.

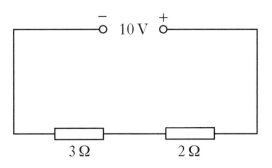

The potential difference across the $2\,\Omega$ resistor is

A   4 V

B   5 V

C   6 V

D   10 V

E   20 V.

8. A student makes the following statements about electrical supplies.

  I   The frequency of the mains supply is 50 Hz.

  II The quoted value of an alternating voltage is less than its peak value.

  III A d.c. supply and an a.c. supply of the same quoted value will supply the same power to a given resistor.

Which of the following statements is/are correct?

A   I only

B   II only

C   III only

D   I and II only

E   I, II and III

[Turn over

9. A wind speed meter is designed as shown.

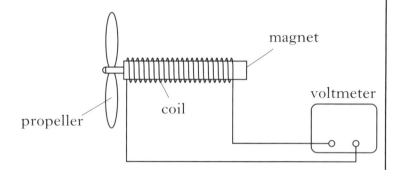

Air blows across the propeller causing the magnet to rotate. A voltage is induced across the coil.

Which of the following changes will produce an increase in the induced voltage?

   I Replacing the magnet with one of greater field strength.

   II Spinning the propeller faster.

   III Reducing the number of turns on the coil.

   A I only

   B I and II only

   C I and III only

   D II and III only

   E I, II and III

10. Which of the following is the circuit symbol for an NPN transistor?

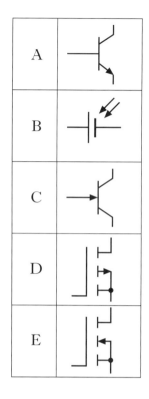

11. The input signal to an amplifier is 2 V a.c. at a frequency of 200 Hz. The amplifier has a gain of 8.

Which row shows the output voltage and the output frequency?

	Output voltage (V)	Output frequency (Hz)
A	10	200
B	10	208
C	10	1600
D	16	200
E	16	1600

12. The following diagram gives information about a wave.

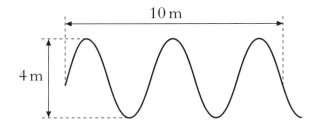

Which row shows the amplitude and wavelength of the wave?

	Amplitude (m)	Wavelength (m)
A	2	2
B	2	4
C	2	5
D	4	2
E	4	4

13. Sound is a longitudinal wave. When a sound wave travels through air the particles of air

A vibrate at random

B vibrate along the wave direction

C vibrate at 90° to the wave direction

D move continuously away from the source

E move continuously towards the source.

14. A signal is transmitted using a curved reflector as shown.

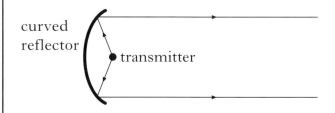

Which of the following statements is/are correct?

I The signal meets the curved reflector at an angle called the critical angle.

II The transmitter is placed at the focus of the reflector.

III At the curved reflector, the angle of reflection of the signal is equal to the angle of incidence.

A I only

B I and II only

C I and III only

D II and III only

E I, II and III

[Turn over

15. The diagram shows a ray of light P incident on a rectangular glass block.

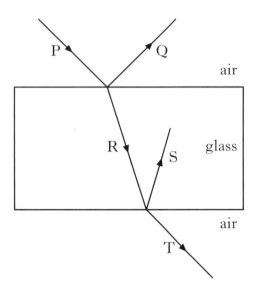

Which of the following are refracted rays?

A   Q and R

B   R and S

C   S and T

D   Q and S

E   R and T

16. The diagram shows the path of a ray of red light in a glass block.

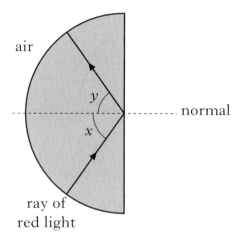

A student makes the following statements.

  I   Angle x is equal to angle y.

 II   Total internal reflection is taking place.

III   Angle x is the critical angle for this glass.

Which of the following statements is/are correct?

A   I only

B   II only

C   I and II only

D   II and III only

E   I, II and III

17. Activity and absorbed dose are quantities used in Dosimetry.

Which row shows the unit of activity and the unit of absorbed dose?

	Unit of activity	Unit of absorbed dose
A	gray	becquerel
B	becquerel	sievert
C	becquerel	gray
D	gray	sievert
E	sievert	gray

18. The table shows the count rate of a radioactive source taken at regular time intervals. The count rate has been corrected for background radiation.

Time (minutes)	10	20	30	40	50
Count rate (counts per minute)	800	630	500	400	315

What is the half-life in minutes of the isotope?

A   10

B   15

C   20

D   30

E   40

19. In the following passage some words have been replaced by letters X and Y.

In a nuclear reactor, fission is caused by X bombardment of a uranium nucleus. This causes the nucleus to split releasing neutrons and Y.

Which row gives the words for X and Y?

	X	Y
A	neutron	energy
B	proton	energy
C	electron	protons
D	neutron	protons
E	electron	energy

20. Control rods in a nuclear reactor

A   absorb neutrons

B   contain uranium

C   produce neutrons

D   remove heat from the reactor

E   slow down neutrons.

Candidates are reminded that the answer sheet for Section A MUST be placed INSIDE the front cover of the answer book.

[Turn over

SECTION B   *Marks*

Write your answers to questions 21–31 in the answer book.

All answers must be written clearly and legibly in ink.

21. A cricketer strikes a ball. The ball leaves the bat horizontally at 20 m/s. It hits the ground at a distance of 11 m from the point where it was struck.

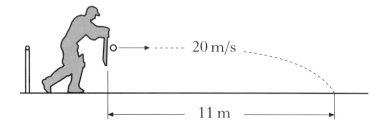

Assume that air resistance is negligible.

(a) Calculate the time of flight of the ball.   2

(b) Calculate the vertical speed of the ball as it reaches the ground.   2

(c) Sketch a graph of vertical speed against time for the ball. Numerical values are required on both axes.   2

(d) Calculate the vertical distance travelled by the ball during its flight.   2

(8)

Marks

22. A satellite moves in a circular orbit around a planet. The satellite travels at a constant speed whilst accelerating.

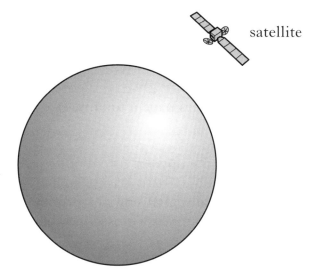

satellite

(a) (i) Define the term *acceleration*. **1**

(ii) Explain how the satellite can be accelerating when it is travelling at a constant speed. **1**

(b) At one particular point in its orbit the satellite fires two rockets. The forces exerted on the satellite by these rockets are shown on the diagram.

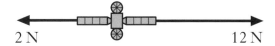

2 N 12 N

The satellite has a mass of 50 kg. Calculate the resultant acceleration due to these forces. **3**

(5)

[Turn over

Marks

23. An aircraft is flying horizontally at a constant speed.

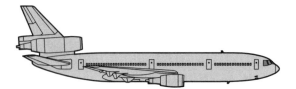

(a) The aircraft and passengers have a total mass of 50 000 kg. Calculate the total weight.   2

(b) State the magnitude of the upward force acting on the aircraft.   1

(c) During the flight, the aircraft's engines produce a force of $4{\cdot}4 \times 10^4$ N due North. The aircraft encounters a crosswind, blowing from west to east, which exerts a force of $3{\cdot}2 \times 10^4$ N.

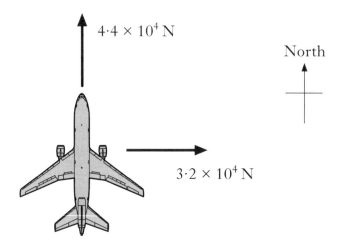

Calculate the resultant force on the aircraft.   3

(d) During a particular flight, a pilot receives an absorbed dose of 15 μGy from gamma rays. Calculate the equivalent dose received due to this type of radiation.   2

(e) Gamma radiation is an example of radiation which causes *ionisation*. Explain what is meant by the term *ionisation*.   1

                                                                                        (9)

Marks

24. An experiment was carried out to determine the specific heat capacity of water. The energy supplied to the water was measured by a joulemeter.

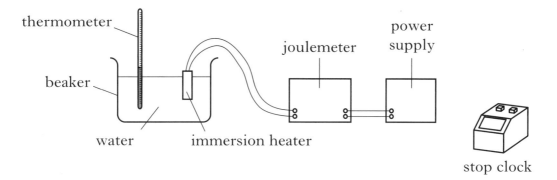

The following data was recorded.

Initial temperature of the water = 21 °C.
Final temperature of the water = 33 °C.
Initial reading on the joulemeter = 12 kJ.
Final reading on the joulemeter = 120 kJ.
Mass of water = 2·0 kg.
Time = 5 minutes.

(a) (i) Calculate the change in temperature of the water.   1

  (ii) Calculate the energy supplied by the immersion heater.   1

  (iii) Calculate the value for the specific heat capacity of water obtained from this experiment.   2

(b) (i) The accepted value for the specific heat capacity of water is quoted in the table in the Data Sheet. Explain the difference between the accepted value and the value obtained in the experiment.   2

  (ii) How could the experiment be improved to reduce this difference?   1

(c) Calculate the power rating of the immersion heater.   2

(9)

[Turn over

Marks

25. Part of a circuit is shown below.

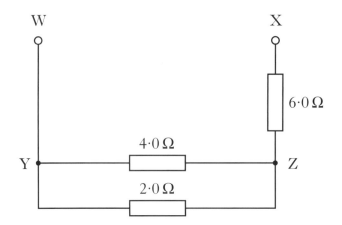

(a) Calculate the total resistance between points Y and Z.   **2**

(b) Calculate the total resistance between points W and X.   **2**

(c) Calculate the voltage across the $2\cdot0\,\Omega$ resistor when the current in the $4\cdot0\,\Omega$ resistor is $0\cdot10\,\text{A}$.   **2**

(6)

Marks

26. A student has two electrical power supplies. One is an a.c. supply and the other is a d.c. supply.

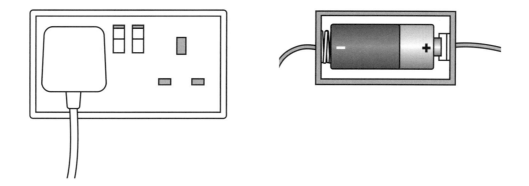

(a) Explain a.c. and d.c. in terms of electron flow in a circuit. 2

(b) The student uses **one** of the supplies to operate a transformer.

   (i) Which power supply should be used to operate the transformer? 1

   (ii) What is the purpose of a transformer? 1

   (iii) A transformer with an efficiency of 30% is used in a computer. Calculate the output power when the input power is 50 W. 2

           (6)

[Turn over

Marks

27. Light emitting diodes (LEDs) are often used as on/off indicators on televisions and computers.

    An LED is connected in a circuit with a resistor R.

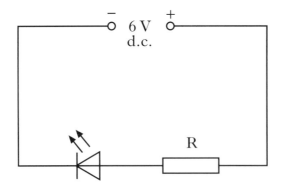

(a) What is the purpose of resistor R? 1

(b) The LED is rated at 2 V, 100 mA. Calculate the resistance of resistor R. 3

(c) Calculate the power developed by resistor R when the LED is working normally. 2

(6)

Marks

28. A solar cell is tested for use in a buggy.

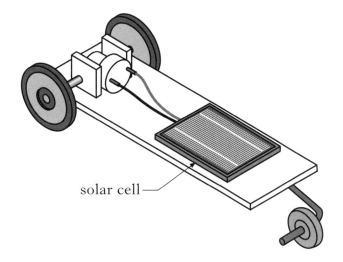

solar cell

The solar cell produces a voltage of 0·5 V and a current of 0·4 mA.

(a) (i) Calculate the power produced by the solar cell. 2

    (ii) The buggy requires 4 mW to operate. Calculate the number of solar cells required to supply this power. 2

(b) State the energy change in a solar cell. 1

(c) The solar cell is illuminated by light of frequency $6·7 \times 10^{14}$ Hz. Calculate the wavelength of this light. 2

(7)

[Turn over

Marks

29. The Sun produces electromagnetic radiation. The electromagnetic spectrum is shown in order of increasing wavelength. Two radiations P and Q have been omitted.

Gamma rays	X rays	P	Visible light	Infra red	Q	Television and radio rays

→ Increasing wavelength

(a) (i) Identify radiations P and Q. 　2

(ii) The planet Neptune is 4.50×10^9 km from the Sun. Calculate the time taken for radio waves from the Sun to reach Neptune. 　2

(iii) State what happens to the frequency of electromagnetic radiation as the wavelength increases. 　1

(b) The Sun produces a *solar wind* consisting of charged particles. In one particular part of the solar wind, a charge of 360 C passes a point in space in one minute. Calculate the current. 　2

(7)

Marks

30. A converging lens has a focal length of 30 mm.

(a) Calculate the power of the lens. **2**

(b) On the graph paper provided, copy and complete the diagram below. Show the size and position of the image formed by the lens.

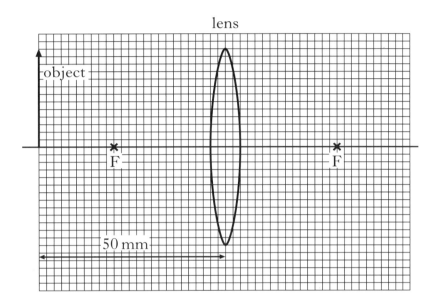

3

(c) Name the eye defect which a converging lens can correct. Explain your answer. **2**

(7)

[Turn over for Question 31 on *Page twenty*

Marks

31. It is possible to determine the age of a prehistoric wooden boat by measuring the activity of radioactive carbon-14.

The activity of a piece of wood from the boat is $300\,\mu$Bq.

(a) How many atoms of carbon-14 decay in 1 day?   2

(b) When the boat was carved, the activity of the piece of wood was $2400\,\mu$Bq due to carbon-14 atoms. The half-life of carbon-14 is 5730 years. Calculate the age of the boat.   2

(c) Carbon-14 emits beta particles. What is a beta particle?   1

(d) A radioactive source emits alpha particles. What is an alpha particle?   1

(e) How does the ionisation density of alpha particles compare with that of beta particles?   1

(f)　(i) A student sets up an experiment as shown.

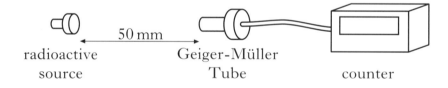

radioactive　　　　　　　　Geiger-Müller　　　　counter
source　　　　　　　　　　　　Tube

The student places a 3 mm sheet of aluminium between the radioactive source and the Geiger-Müller Tube. The count rate is observed to decrease and the student concludes that the radioactive material is emitting beta radiation.

Suggest **one** reason why her conclusion may be incorrect.   1

　　(ii) State **two** safety precautions that the student must observe when handling radioactive sources.   2

(10)

[END OF QUESTION PAPER]

[BLANK PAGE]

X069/11/02

NATIONAL QUALIFICATIONS 2012	MONDAY, 28 MAY 1.00 PM – 3.00 PM	PHYSICS INTERMEDIATE 2

Read Carefully

Reference may be made to the Physics Data Booklet

1   All questions should be attempted.

Section A (questions 1 to 20)

2   Check that the answer sheet is for Physics Intermediate 2 (Section A).

3   For this section of the examination you must use an **HB pencil** and, where necessary, an eraser.

4   Check that the answer sheet you have been given has **your name**, **date of birth**, **SCN** (Scottish Candidate Number) and **Centre Name** printed on it.

Do not change any of these details.

5   If any of this information is wrong, tell the Invigilator immediately.

6   If this information is correct, **print** your name and seat number in the boxes provided.

7   There is **only one correct** answer to each question.

8   Any rough working should be done on the question paper or the rough working sheet, **not** on your answer sheet.

9   At the end of the exam, **put the answer sheet for Section A inside the front cover of your answer book**.

10   Instructions as to how to record your answers to questions 1–20 are given on page three.

Section B (questions 21 to 30)

11   Answer the questions numbered 21 to 30 in the answer book provided.

12   **All answers must be written clearly and legibly in ink.**

13   Fill in the details on the front of the answer book.

14   Enter the question number clearly in the margin of the answer book beside each of your answers to questions 21 to 30.

15   Care should be taken to give an appropriate number of significant figures in the final answers to calculations.

DATA SHEET

Speed of light in materials

Material	Speed in m/s
Air	$3 \cdot 0 \times 10^8$
Carbon dioxide	$3 \cdot 0 \times 10^8$
Diamond	$1 \cdot 2 \times 10^8$
Glass	$2 \cdot 0 \times 10^8$
Glycerol	$2 \cdot 1 \times 10^8$
Water	$2 \cdot 3 \times 10^8$

Speed of sound in materials

Material	Speed in m/s
Aluminium	5200
Air	340
Bone	4100
Carbon dioxide	270
Glycerol	1900
Muscle	1600
Steel	5200
Tissue	1500
Water	1500

Gravitational field strengths

	Gravitational field strength on the surface in N/kg
Earth	10
Jupiter	26
Mars	4
Mercury	4
Moon	$1 \cdot 6$
Neptune	12
Saturn	11
Sun	270
Venus	9

Specific heat capacity of materials

Material	Specific heat capacity in J/kg °C
Alcohol	2350
Aluminium	902
Copper	386
Glass	500
Ice	2100
Iron	480
Lead	128
Oil	2130
Water	4180

Specific latent heat of fusion of materials

Material	Specific latent heat of fusion in J/kg
Alcohol	$0 \cdot 99 \times 10^5$
Aluminium	$3 \cdot 95 \times 10^5$
Carbon Dioxide	$1 \cdot 80 \times 10^5$
Copper	$2 \cdot 05 \times 10^5$
Iron	$2 \cdot 67 \times 10^5$
Lead	$0 \cdot 25 \times 10^5$
Water	$3 \cdot 34 \times 10^5$

Melting and boiling points of materials

Material	Melting point in °C	Boiling point in °C
Alcohol	−98	65
Aluminium	660	2470
Copper	1077	2567
Glycerol	18	290
Lead	328	1737
Iron	1537	2737

Specific latent heat of vaporisation of materials

Material	Specific latent heat of vaporisation in J/kg
Alcohol	$11 \cdot 2 \times 10^5$
Carbon Dioxide	$3 \cdot 77 \times 10^5$
Glycerol	$8 \cdot 30 \times 10^5$
Turpentine	$2 \cdot 90 \times 10^5$
Water	$22 \cdot 6 \times 10^5$

Radiation weighting factors

Type of radiation	Radiation weighting factor
alpha	20
beta	1
fast neutrons	10
gamma	1
slow neutrons	3

SECTION A

For questions 1 to 20 in this section of the paper the answer to each question is either A, B, C, D or E. Decide what your answer is, then, using your pencil, put a horizontal line in the space provided—see the example below.

EXAMPLE

The energy unit measured by the electricity meter in your home is the

   A kilowatt-hour

   B ampere

   C watt

   D coulomb

   E volt.

The correct answer is **A**—kilowatt-hour. The answer **A** has been clearly marked in **pencil** with a horizontal line (see below).

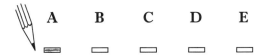

Changing an answer

If you decide to change your answer, carefully erase your first answer and, using your pencil, fill in the answer you want. The answer below has been changed to **E**.

[Turn over

SECTION A

Answer questions 1–20 on the answer sheet.

1. At an airport an aircraft moves from the terminal building to the end of the runway.

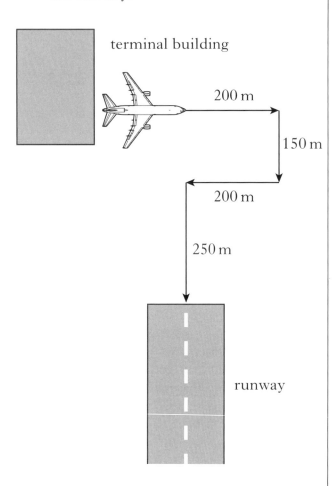

Which row shows the total distance travelled and the size of the displacement of the aircraft?

	Total distance travelled (m)	Size of displacement (m)
A	400	800
B	450	200
C	450	400
D	800	400
E	800	800

2. Near the Earth's surface, a mass of 6 kg is falling with a constant velocity.

The air resistance and the unbalanced force acting on the mass are:

	air resistance	unbalanced force
A	60 N upwards	0 N
B	10 N upwards	10 N downwards
C	10 N downwards	70 N downwards
D	10 N upwards	0 N
E	60 N upwards	60 N downwards

3. Two forces act on an object O in the directions shown.

The size of the resultant force is

A 14 N

B 24 N

C 38 N

D 45 N

E 62 N.

4. The diagram shows the horizontal forces acting on a box.

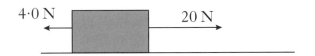

The box accelerates at $1 \cdot 6 \, \text{m/s}^2$.

The mass of the box is

A $0 \cdot 10 \, \text{kg}$

B $10 \cdot 0 \, \text{kg}$

C $15 \cdot 0 \, \text{kg}$

D $25 \cdot 6 \, \text{kg}$

E $38 \cdot 4 \, \text{kg}$.

5. Two identical balls X and Y are projected horizontally from the edge of a cliff.

The path taken by each ball is shown.

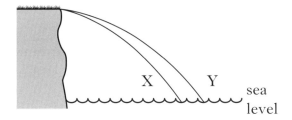

A student makes the following statements about the motion of the two balls.

I   They take the same time to reach sea level.

II   They have the same vertical acceleration.

III They have the same horizontal velocity.

Which of these statements is/are correct?

A   I only

B   II only

C   I and II only

D   I and III only

E   II and III only

[Turn over

6. Car X of mass 1500 kg travels at 20 m/s along a straight, horizontal road. It collides with a stationary car Y of mass 1900 kg.

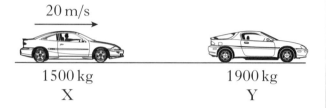

20 m/s

1500 kg   1900 kg
X   Y

The two cars lock together after the collision.

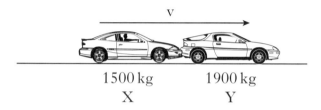

v

1500 kg   1900 kg
X   Y

The speed of the cars after the collision is

A   8·8 m/s

B   9·4 m/s

C   11 m/s

D   16 m/s

E   20 m/s.

7. An electrical motor raises a crate of mass 500 kg through a height of 12 m in 4 s.

The minimum power rating of the motor is

A   1·25 kW

B   1·5 kW

C   15 kW

D   60 kW

E   240 kW.

8. A heater is immersed in a substance. The heater is then switched on.

The graph shows the temperature of the substance over a period of time.

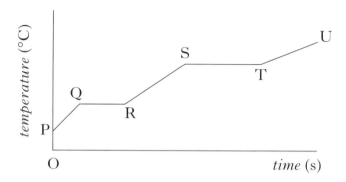

Which row in the table identifies the sections of the graph when the substance is changing state?

	Solid to liquid	Liquid to gas
A	QR	TU
B	QR	ST
C	PQ	RS
D	PQ	TU
E	ST	QR

9. Which row in the table gives the accepted values for the UK mains supply?

	Frequency (Hz)	Quoted voltage (V)	Peak voltage (V)
A	10	110	230
B	50	230	230
C	50	230	325
D	60	230	162
E	230	50	50

10. A circuit contains an ideal transformer connected to a 10 V d.c. supply.

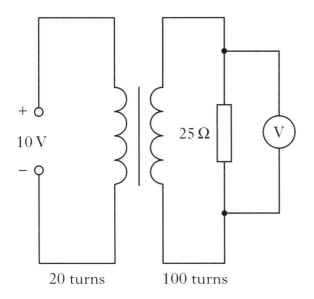

20 turns   100 turns

The potential difference across the 25 Ω resistor is

A   0 V

B   2 V

C   10 V

D   50 V

E   80 V.

11. A student sets up the circuits shown.

In which circuit will both LEDs be lit?

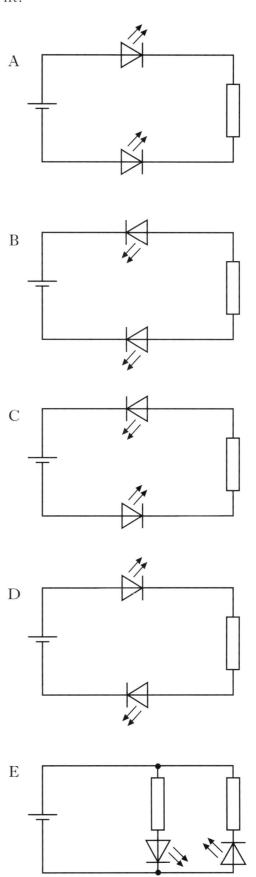

12. Which row in the table correctly identifies input and output devices?

	Input device	Output devices
A	microphone	loudspeaker, LED
B	solar cell	thermocouple, LED
C	loudspeaker	microphone, relay
D	LED	loudspeaker, relay
E	thermocouple	microphone, LED

13. A circuit is set up to test electrical conduction in materials.

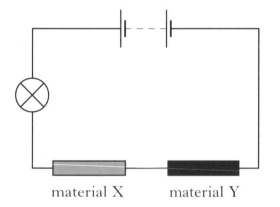

material X   material Y

The lamp lights.

Which row in the table identifies materials X and Y?

	Material X	Material Y
A	copper	wood
B	copper	aluminium
C	glass	copper
D	aluminium	glass
E	wood	glass

14. The current in an 8 Ω resistor is 2 A.

The charge passing through the resistor in 10 s is

A   4 C

B   5 C

C   16 C

D   20 C

E   80 C.

15. Which of the following statements is/are correct?

I   In an a.c. circuit the direction of the current changes regularly.

II   In a d.c. circuit positive charges flow in one direction only.

III In an a.c. circuit the size of the current varies with time.

A   I only

B   II only

C   I and II only

D   I and III only

E   I, II and III

16. A signal of voltage 5·0 mV and frequency 2000 Hz is applied to the input of an amplifier.

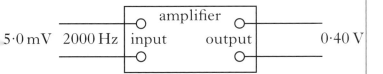

The output voltage is 0·40 V.

Which row shows the voltage gain of the amplifier and the frequency of the output signal?

	Voltage gain	Frequency of output signal (Hz)
A	0·0125	2000
B	0·08	50
C	0·08	2000
D	80	50
E	80	2000

17. The diagram shows two rays of red light X and Y passing through a block of glass.

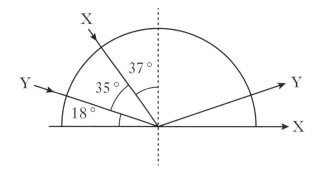

The critical angle of the glass for this light is

A 18°

B 35°

C 37°

D 53°

E 72°.

18. A student makes the following statements.

  I In an atom there are neutrons and electrons in the nucleus and protons which orbit the nucleus.

  II An alpha particle consists of two neutrons and two electrons.

  III A beta particle is a fast moving electron.

Which of the statements is/are correct?

A I only

B II only

C III only

D I and III only

E I, II and III

[Turn over

19. A radioactive source emits alpha, beta and gamma radiation. A detector, connected to a counter, is placed 10 mm in front of the source. The counter records 400 counts per minute.

A sheet of paper is placed between the source and the detector. The counter records 300 counts per minute.

The radiation now detected is

A   alpha only

B   beta only

C   gamma only

D   alpha and beta only

E   beta and gamma only.

20. A radioactive tracer is injected into a patient to study the flow of blood.

The tracer should have a

A   short half-life and emit α particles

B   long half-life and emit β particles

C   long half-life and emit γ rays

D   long half-life and emit α particles

E   short half-life and emit γ rays.

Candidates are reminded that the answer sheet for Section A MUST be placed INSIDE the front cover of the answer book.

[Turn over for Section B on *Page twelve*

SECTION B   *Marks*

Write your answers to questions 21–30 in the answer book.

All answers must be written clearly and legibly in ink.

21. Sputnik 1, the first man-made satellite, was launched in 1957. It orbited the Earth at a speed of 8300 m/s and had a mass of 84 kg.

 (*a*)   (i) Sputnik 1 orbited Earth in 100 minutes.

        Calculate the distance it travelled in this time.   2

   (ii) Although Sputnik 1 travelled at a constant speed in a circular orbit, it accelerated continuously.

        Explain this statement.   2

 (*b*) Sputnik 1 transmitted radio signals a distance of 800 km to the surface of the Earth.

    Calculate the time taken for the signals to reach the Earth's surface.   2

21. (continued)   *Marks*

(c) The graph shows how gravitational field strength varies with height above the surface of the Earth.

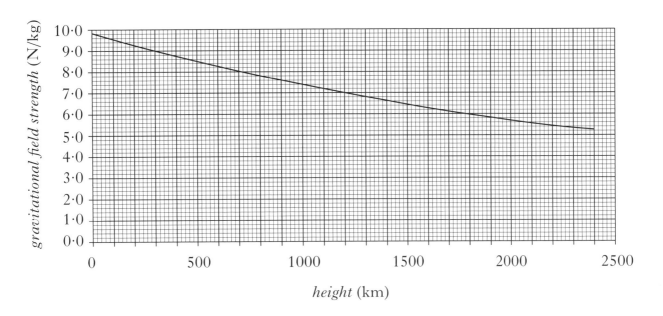

height (km)

(i) Define the term **gravitational field strength**.   1

(ii) What is the value of the gravitational field strength at a height of 800 km?   1

(iii) Calculate the weight of Sputnik 1 at this height.   2

(10)

[Turn over

Marks

22. A car of mass 700 kg travels along a motorway at a constant speed. The driver sees a traffic hold-up ahead and performs an emergency stop. A graph of the car's motion is shown, from the moment the driver sees the hold-up.

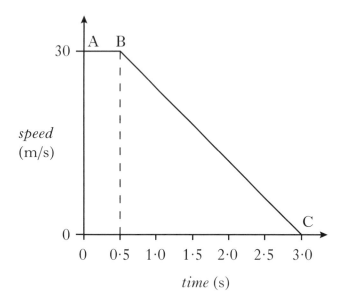

(*a*) Describe **and** explain the motion of the car between A and B.   2

(*b*) Calculate the kinetic energy of the car at A.   2

(*c*) State the work done in bringing the car to a halt between B and C.   1

(*d*) Show by calculation that the magnitude of the unbalanced force required to bring the car to a halt between B and C is 8400 N.   2

(7)

Marks

23. A student reproduces Galilleo's famous experiment by dropping a solid copper ball of mass 0·50 kg from a balcony on the Leaning Tower of Pisa.

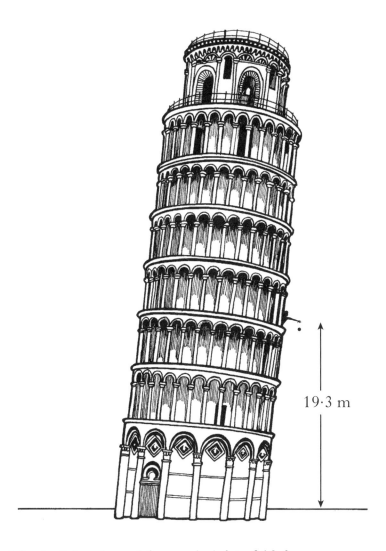

19·3 m

(*a*) (i) The ball is released from a height of 19·3 m.

   Calculate the gravitational potential energy lost by the ball. **2**

   (ii) Assuming that all of this gravitational potential energy is converted into heat energy **in the ball**, calculate the increase in the temperature of the ball on impact with the ground. **2**

   (iii) Is the actual temperature change of the ball greater than, the same as or less than the value calculated in part (*a*)(ii)?

   You **must** explain your answer. **2**

(*b*) The ball was made by melting 0·50 kg of copper at its melting point. Calculate the amount of heat energy required for this. **3**

(9)

Marks

24. A resistor is labelled: "10 Ω ± 10%, 3 W".

This means that the resistance value could actually be between 9 Ω and 11 Ω.

(a) A student decides to check the value of the resistance.

Draw a circuit diagram, including a 6 V battery, a voltmeter and an ammeter, for a circuit that could be used to determine the resistance.   **3**

(b) Readings from the circuit give the voltage across the resistor as 5·7 V and the current in the resistor as 0·60 A.

Use these values to calculate the resistance.   **2**

(c) During this experiment, the resistor becomes very hot and gives off smoke.

Explain why this happens.

You **must** include a calculation as part of your answer.   **3**

(d) The student states that **two** of these resistors would not have overheated if they were connected together in parallel with the battery.

Is the student correct?

Explain your answer.   **2**

(10)

Marks

25. The circuit shown switches a warning lamp on or off depending on the temperature.

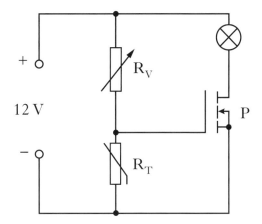

(a) Name component P.

1

(b) As the temperature increases the resistance of thermistor R_T decreases. What happens to the voltage across R_T as the temperature increases?

1

(c) When the voltage applied to component P is equal to or greater than 2·4 V, component P switches on and the warning lamp lights.

R_V is adjusted until its resistance is 5600 Ω and the warning lamp now lights.

At this point calculate:

  (i) the voltage across R_V;

1

  (ii) the resistance of R_T.

2

(d) The temperature of R_T now decreases.

Will the lamp stay on or go off?

You **must** explain your answer.

2

(7)

[Turn over

Marks

26. An apparatus used to measure the speed of sound consists of a bright LED which flashes every 0·5 s and a loudspeaker which beeps at **exactly the same time** as the LED flashes.

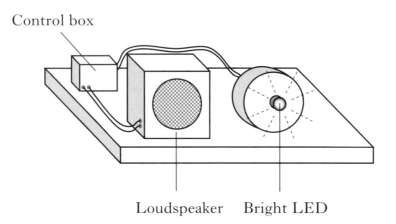

Control box

Loudspeaker   Bright LED

(a) A student standing beside the apparatus observes the beeps and flashes happening at exactly the same time.

Another student standing 88 m away does **not** observe them happening at the same time.

   (i) Explain this observation. **1**

   (ii) A third student 176 m away observes the beeps and flashes happening at exactly the same time.

      Use this information to calculate a value for the speed of sound. **2**

(b) The circuit used to operate the LED and loudspeaker uses electric switches called relays.

Simplified diagrams of a relay are shown.

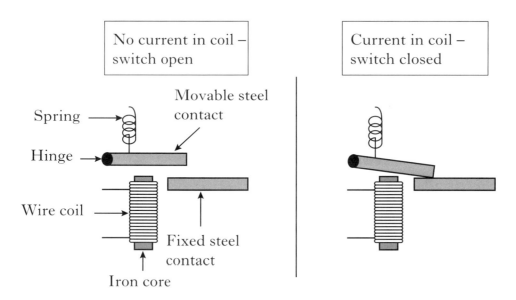

No current in coil – switch open

Current in coil – switch closed

Spring
Movable steel contact
Hinge
Wire coil
Fixed steel contact
Iron core

   Explain why the steel contact moves when there is a current in the coil. **2**

26. (continued)   *Marks*

(*c*)   Part of the circuit for the apparatus is shown.

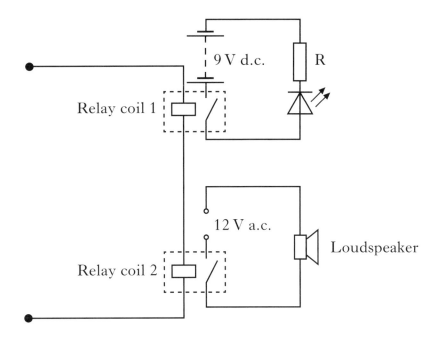

The LED is rated 6·0 V, 800 mA.

Calculate the value of resistor R.   **3**

(*d*)   The 12 V a.c. supply has a frequency of 850 Hz.

Using the value of the speed of sound from the data sheet, calculate the wavelength of the sound in air produced by the loudspeaker.   **2**

(10)

[Turn over

Marks

27. Optical fibres are used to carry internet data using infra-red radiation.

    (a) Is the wavelength of infra-red radiation greater than, the same as, or less than the wavelength of visible light?   **1**

    (b) The diagram shows a view of an infra-red ray entering the end of a fibre.

    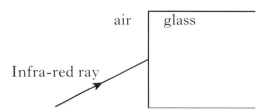

    Copy and complete the diagram to show the path of the infra-red ray as it enters the glass from air.

    Indicate on your diagram the normal, the angle of incidence and the angle of refraction.   **2**

    (c) The diagram shows the path of the infra-red ray as it passes through a section of the fibre.

    Name the effect shown.   **1**

    (4)

[Turn over for Question 28 on *Page twenty-two*

Marks

28. The picture shows a spy using a long range microphone and curved reflector to listen to conversations from some distance away.

(*a*) Copy and complete the following diagram to show the path of the sound waves.

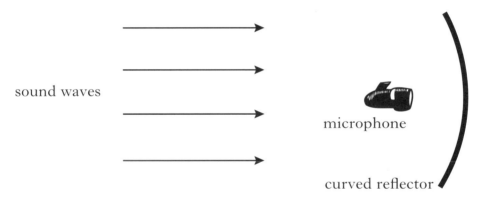

2

(*b*) Explain why using the curved reflector makes the sound detected by the microphone louder.

1

(*c*) The microphone produces a signal of 24 mV which is applied to the input of an amplifier.

The output from the amplifier is 3·0 V.

Calculate the voltage gain of the amplifier.

2

Marks

28. (continued)

(*d*) The spy needs spectacles to see distant objects clearly.

    (i) What is the name given to this eye defect?   **1**

    (ii) What type of lens is needed to correct this defect?   **1**

(*e*) The spy uses a magnifying lens of power +10 D to examine some photographs.

Calculate the focal length of this lens.   **2**

    (9)

[Turn over

Marks

29. A technician checks the count rate of a radioactive source. A graph of count rate against time for the source is shown. The count rate has been corrected for background radiation.

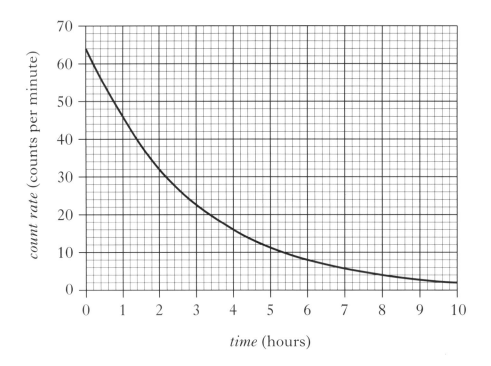

time (hours)

(a) Use the graph to determine the half-life of the source.   2

(b) State **two** factors which can affect the background radiation level.   2

(c) The source emits gamma rays. State what is meant by a gamma ray.   1

(5)

Marks

30. An ageing nuclear power station is being dismantled.

(*a*) During the dismantling process a worker comes into contact with an object that emits 24 000 alpha particles in five minutes. The worker's hand has a mass of 0·50 kg and absorbs 6·0 μJ of energy.

Calculate:

(i) the absorbed dose received by the worker's hand;   2

(ii) the equivalent dose received by the worker's hand;   2

(iii) the activity of the object.   2

(*b*) In a nuclear reactor, state the purpose of:

(i) the moderator;   1

(ii) the containment vessel.   1

(*c*) What type of nuclear reaction takes place in a nuclear power station's reactor?   1

    (9)

[END OF QUESTION PAPER]

[BLANK PAGE]

INTERMEDIATE 2

2013

[BLANK PAGE]

X069/11/02

NATIONAL
QUALIFICATIONS
2013

MONDAY, 27 MAY
1.00 PM – 3.00 PM

PHYSICS
INTERMEDIATE 2

Read Carefully

Reference may be made to the Physics Data Booklet

1   All questions should be attempted.

Section A (questions 1 to 20)

2   Check that the answer sheet is for Physics Intermediate 2 (Section A).

3   For this section of the examination you must use an **HB pencil** and, where necessary, an eraser.

4   Check that the answer sheet you have been given has **your name**, **date of birth**, **SCN** (Scottish Candidate Number) and **Centre Name** printed on it.

    Do not change any of these details.

5   If any of this information is wrong, tell the Invigilator immediately.

6   If this information is correct, **print** your name and seat number in the boxes provided.

7   There is **only one correct** answer to each question.

8   Any rough working should be done on the question paper or the rough working sheet, **not** on your answer sheet.

9   At the end of the exam, **put the answer sheet for Section A inside the front cover of your answer book**.

10   Instructions as to how to record your answers to questions 1–20 are given on page three.

Section B (questions 21 to 31)

11   Answer the questions numbered 21 to 31 in the answer book provided.

12   **All answers must be written clearly and legibly in ink.**

13   Fill in the details on the front of the answer book.

14   Enter the question number clearly in the margin of the answer book beside each of your answers to questions 21 to 31.

15   Care should be taken to give an appropriate number of significant figures in the final answers to calculations.

DATA SHEET

Speed of light in materials

Material	Speed in m/s
Air	$3 \cdot 0 \times 10^8$
Carbon dioxide	$3 \cdot 0 \times 10^8$
Diamond	$1 \cdot 2 \times 10^8$
Glass	$2 \cdot 0 \times 10^8$
Glycerol	$2 \cdot 1 \times 10^8$
Water	$2 \cdot 3 \times 10^8$

Speed of sound in materials

Material	Speed in m/s
Aluminium	5200
Air	340
Bone	4100
Carbon dioxide	270
Glycerol	1900
Muscle	1600
Steel	5200
Tissue	1500
Water	1500

Gravitational field strengths

	Gravitational field strength on the surface in N/kg
Earth	10
Jupiter	26
Mars	4
Mercury	4
Moon	$1 \cdot 6$
Neptune	12
Saturn	11
Sun	270
Venus	9

Specific heat capacity of materials

Material	Specific heat capacity in J/kg °C
Alcohol	2350
Aluminium	902
Copper	386
Glass	500
Ice	2100
Iron	480
Lead	128
Oil	2130
Water	4180

Specific latent heat of fusion of materials

Material	Specific latent heat of fusion in J/kg
Alcohol	$0 \cdot 99 \times 10^5$
Aluminium	$3 \cdot 95 \times 10^5$
Carbon Dioxide	$1 \cdot 80 \times 10^5$
Copper	$2 \cdot 05 \times 10^5$
Iron	$2 \cdot 67 \times 10^5$
Lead	$0 \cdot 25 \times 10^5$
Water	$3 \cdot 34 \times 10^5$

Melting and boiling points of materials

Material	Melting point in °C	Boiling point in °C
Alcohol	−98	65
Aluminium	660	2470
Copper	1077	2567
Glycerol	18	290
Lead	328	1737
Iron	1537	2737

Specific latent heat of vaporisation of materials

Material	Specific latent heat of vaporisation in J/kg
Alcohol	$11 \cdot 2 \times 10^5$
Carbon Dioxide	$3 \cdot 77 \times 10^5$
Glycerol	$8 \cdot 30 \times 10^5$
Turpentine	$2 \cdot 90 \times 10^5$
Water	$22 \cdot 6 \times 10^5$

Radiation weighting factors

Type of radiation	Radiation weighting factor
alpha	20
beta	1
fast neutrons	10
gamma	1
slow neutrons	3

SECTION A

For questions **1 to 20** in this section of the paper the answer to each question is either **A, B, C, D** or **E**. Decide what your answer is, then, using your pencil, put a horizontal line in the space provided—see the example below.

EXAMPLE

The energy unit measured by the electricity meter in your home is the

    A   kilowatt-hour

    B   ampere

    C   watt

    D   coulomb

    E   volt.

The correct answer is **A**—kilowatt-hour. The answer **A** has been clearly marked in **pencil** with a horizontal line (see below).

Changing an answer

If you decide to change your answer, carefully erase your first answer and, using your pencil, fill in the answer you want. The answer below has been changed to **E**.

    A   B   C   D   E

[Turn over

SECTION A

Answer questions 1–20 on the answer sheet.

1. Which row contains two scalar quantities and one vector quantity?

   A   Distance, momentum, velocity

   B   Speed, mass, momentum

   C   Distance, weight, force

   D   Speed, weight, momentum

   E   Velocity, force, mass

2. A student follows the route shown in the diagram and arrives back at the starting point.

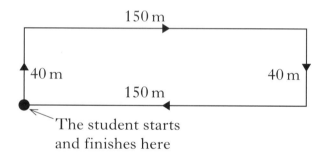

   Which row in the table shows the total distance walked and the magnitude of the final displacement?

	Total distance (m)	Final displacement (m)
A	0	80
B	0	380
C	190	0
D	380	0
E	380	380

3. A space probe has a mass of 60 kg. The weight of the space probe at the surface of a planet in our solar system is 720 N.

   The planet is

   A   Venus

   B   Mars

   C   Jupiter

   D   Saturn

   E   Neptune.

4. A block is pulled across a horizontal surface as shown.

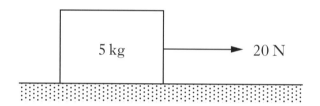

   The mass of the block is 5 kg.

   The block is travelling at a constant speed.

   The force of friction acting on the block is

   A   0 N

   B   4 N

   C   15 N

   D   20 N

   E   25 N.

5. Four tugs apply forces to an oil-rig in the directions shown.

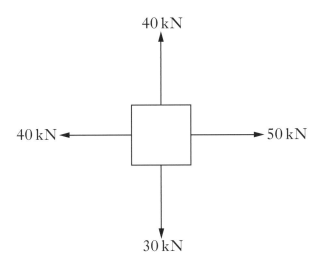

Which of the following could represent the direction of the resultant force?

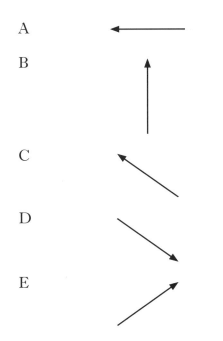

6. The specific latent heat of fusion of a substance is the energy required to

A melt 1 kg of the substance at its melting point

B evaporate 1 kg of the substance at its boiling point

C change the state of the substance without changing its temperature

D change the temperature of the substance without changing its state

E change the temperature of 1 kg of the substance by 1 °C.

7. A block of ice of mass 1·5 kg is placed in a room.

The temperature of the block is 0 °C.

The temperature of the room is 20 °C.

The minimum energy required to **melt** the ice is

A $0·63 \times 10^5$ J

B $1·25 \times 10^5$ J

C $1·88 \times 10^5$ J

D $5·01 \times 10^5$ J

E $6·26 \times 10^5$ J.

[Turn over

8. A circuit with three gaps is shown below.

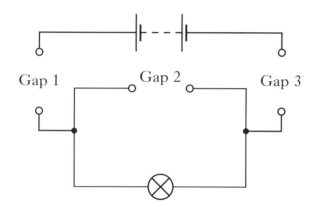

Gap 1   Gap 2   Gap 3

Which row in the table shows the combination of conductors and insulators that should be placed in the gaps to allow the lamp to light?

	Gap 1	Gap 2	Gap 3
A	conductor	insulator	conductor
B	conductor	conductor	insulator
C	conductor	conductor	conductor
D	insulator	insulator	conductor
E	insulator	insulator	insulator

9. In which circuit below would the meter readings allow the resistance of R_2 to be calculated?

A

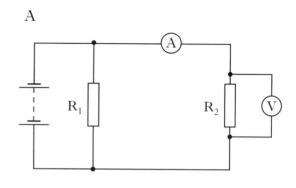

B

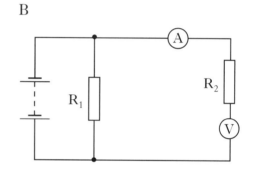

C

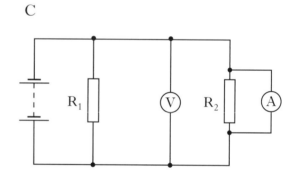

D

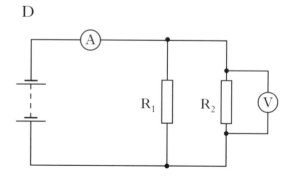

E

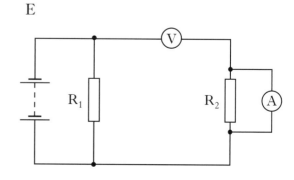

10. A circuit is set up as shown.

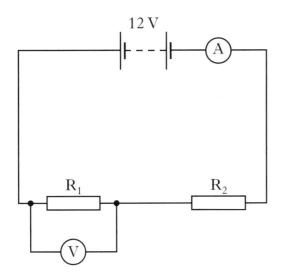

The reading on the ammeter is 3·0 A.

The reading on the voltmeter is 4·0 V.

Which row in the table shows the current in resistor R_2 and the voltage across resistor R_2?

	Current in resistor R_2 (A)	Voltage across resistor R_2 (V)
A	1·5	8·0
B	3·0	4·0
C	3·0	8·0
D	1·5	12·0
E	6·0	4·0

11. A circuit is set up as shown.

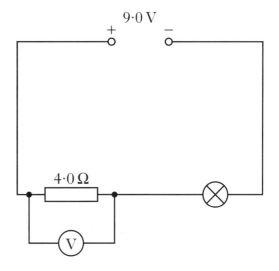

The current in the lamp is 1·5 A.

The reading on the voltmeter is 6·0 V.

The power developed in the lamp is

A   3·0 W

B   4·5 W

C   6·0 W

D   9·0 W

E   13·5 W.

12. Which of the following devices converts heat energy into electrical energy?

A   Solar cell

B   Resistor

C   Thermocouple

D   Thermistor

E   Transistor

[Turn over

13. Which of the following electromagnetic waves has a higher frequency than microwaves and a lower frequency than visible light?

A   Gamma rays

B   Infrared

C   Radio

D   Ultraviolet

E   X-rays

14. A ray of light is incident on a plane mirror as shown.

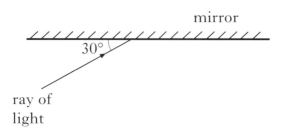

The ray of light reflects from the mirror.

Which row in the table shows the values of the angle of incidence and the angle of reflection?

	Angle of incidence	Angle of reflection
A	30°	30°
B	30°	60°
C	30°	150°
D	60°	30°
E	60°	60°

15. The diagram shows a ray of light in an optical fibre.

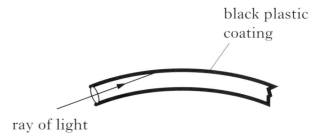

A student makes the following statements about light transmitted along the optical fibre.

  I   The light is totally internally reflected inside the glass.

  II The light is reflected by the black plastic coating.

  III The angle of incidence in the glass is greater than the critical angle for this glass.

Which of the statements is/are correct?

A   I only

B   III only

C   I and II only

D   I and III only

E   I, II and III

16. Which of the following diagrams shows the focusing of rays of light from a distant object by the eye of a long-sighted person?

A

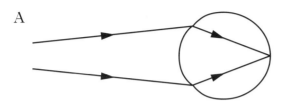

B

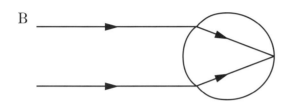

C

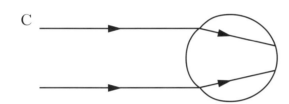

D

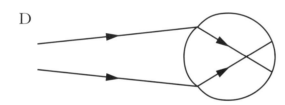

E

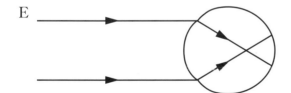

17. A student makes the following statements.

  I   The nucleus of an atom contains protons and electrons.

  II   Gamma radiation produces the greatest ionisation density.

  III   Beta particles are fast moving electrons.

Which of the statements is/are correct?

A   I only

B   II only

C   III only

D   I and III only

E   II and III only

[Turn over

18. A radioactive source emits α, β and γ radiation.

Sheets of aluminium and paper are placed close to the source as shown.

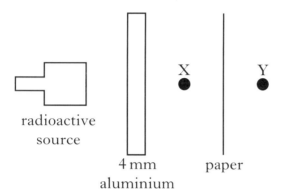

radioactive source

4 mm aluminium   paper

Which row in the table shows the radiation(s) from the source detected at points X and Y?

	Radiation(s) detected at X	Radiation detected at Y
A	α, γ	γ
B	β, γ	α
C	α	β
D	β	γ
E	γ	γ

19. Which of the following describes the term ionisation?

A   An atom losing an orbiting electron.

B   An atom losing a proton.

C   A nucleus emitting an alpha particle.

D   A nucleus emitting a neutron.

E   A nucleus emitting a gamma ray.

20. A student makes the following statements about radiation.

I   The half life of a radioactive source is half of the time it takes for its activity to reduce to zero.

II   The activity of a radioactive source is the number of decays per minute.

III   The risk of biological harm from radiation depends on the type of tissue exposed.

Which of the statements is/are correct?

A   I only

B   II only

C   III only

D   II and III only

E   I, II and III

Candidates are reminded that the answer sheet for Section A MUST be placed INSIDE the front cover of the answer book.

SECTION B

Marks

Write your answers to questions 21–31 in the answer book.

All answers must be written clearly and legibly in ink.

21. A plane of mass 750 kg is at rest on a runway. The engine applies a force of 4·50 kN.

(a) Calculate the magnitude of the acceleration of the plane assuming there are no other forces acting on the plane at this point.

2

(b) The required speed for take-off is 54 m/s.

Calculate the time it takes to reach this speed assuming the acceleration is constant.

2

(c) In practice the acceleration is not constant. Give a reason for this.

1

(5)

[Turn over

22. A student uses a linear air track and an ultrasonic motion sensor to investigate a collision between two vehicles.

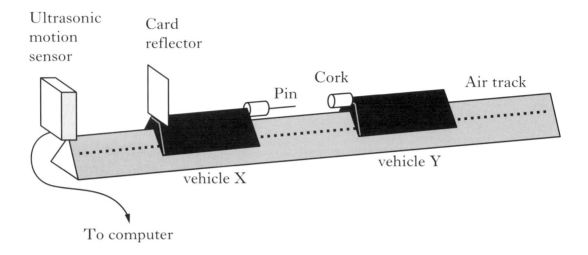

Vehicle Y is at rest before the collision.

Vehicle X is given a push and then released.

A pin on vehicle X sticks into a cork on vehicle Y causing them to join and move off together.

The motion sensor measures the speed of vehicle X every 0·01 s.

The graph shows the results obtained from the investigation after vehicle X has been released.

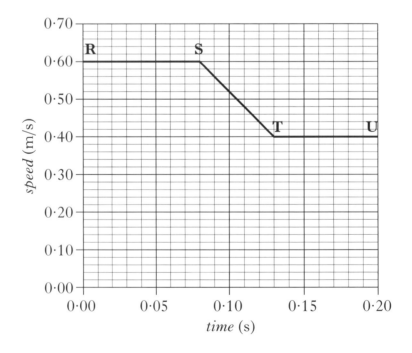

Marks

22. **(continued)**

(*a*) State the speed of ultrasonic waves in air.   **1**

(*b*)   (i) Describe the motion of vehicle X between points **S** and **T**.   **1**

(ii) Calculate the distance travelled by vehicle X between points **S** and **T**.   **2**

(iii) Vehicle X has a mass of 0·50 kg.

Use the law of conservation of momentum to show that vehicle Y has a mass of 0·25 kg.   **2**

(iv) (A)   Calculate the kinetic energy lost in this collision.   **3**

(B)   What happens to the lost kinetic energy?   **1**

(10)

[Turn over

Marks

23. In a TV game show contestants are challenged to run off a horizontal platform and land in a rubber ring floating in a swimming pool.

The platform is 2·8 m above the water surface.

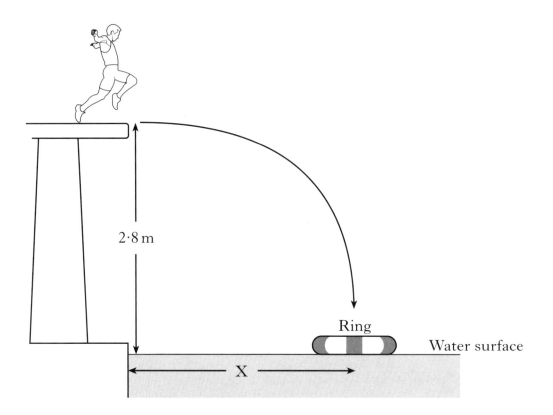

   (*a*) A contestant has a mass of 60 kg.

   He runs off the platform with a horizontal velocity of 2 m/s. He takes 0·75 s to reach the water surface in the centre of the ring.

   (i) Calculate the horizontal distance X from the poolside to the centre of the ring. **2**

   (ii) Calculate the vertical velocity of the contestant as he reaches the water surface. **2**

   (*b*) Another contestant has a mass of 80 kg.

   Will she need to run faster, slower or at the same horizontal speed as the first contestant to land in the ring?

   You **must** explain your answer. **2**

   (6)

Marks

24.   In a garage, a mechanic lifts an engine from a car using a pulley system.

(*a*)   The mechanic pulls 4·5 m of chain with a constant force of 250 N.

Calculate the work done by the mechanic.   2

(*b*)   The engine has a mass of 144 kg and is raised 0·75 m.

Calculate the gravitational potential energy gained by the engine.   2

(*c*)   Calculate the percentage efficiency of the pulley system.   2

(6)

[Turn over

Marks

25. The rating plate on a microwave oven shows the following data.

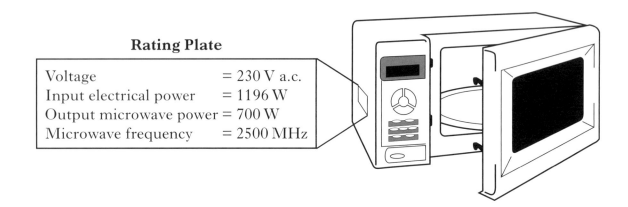

Rating Plate

Voltage	= 230 V a.c.
Input electrical power	= 1196 W
Output microwave power	= 700 W
Microwave frequency	= 2500 MHz

(a) State what is meant by the term voltage.   1

(b)   (i) Calculate the input current.   2

   (ii) The microwave is used to heat a cup of milk for 1 minute 30 seconds. Calculate how much electrical charge passes through the flex in this time.   2

   (iii) The milk of mass 0·25 kg absorbs 48 kJ of energy during the heating process. The specific heat capacity of milk is 3900 J/kg °C.

      Calculate the temperature rise in the milk.   2

(c) Calculate the wavelength of the microwaves.   2

                               (9)

[Turn over for Question 26 on *Page eighteen*

Marks

26. An overhead projector contains a lamp and a motor that operates a cooling fan.

    A technician has a choice of two lamps to fit in the projector.

Lamp A: rated 24·0 V, 2·5 Ω

Lamp B: rated 24·0 V, 5·4 Ω

(*a*) Which lamp gives a brighter light when operating at the correct voltage?
Explain your answer.   2

(*b*) Calculate the power developed by lamp A when it is operating normally.   2

(*c*) The overhead projector plug contains a fuse.
Draw the circuit symbol for a fuse.   1

(*d*) The technician builds a test circuit containing a resistor and a motor, as shown
in **Circuit 1**.

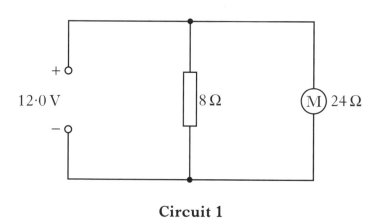

Circuit 1

    (i) State the voltage across the motor.   1

    (ii) Calculate the combined resistance of the resistor and the motor.   2

Marks

26. (continued)

(*e*) The resistor and the motor are now connected in series, as shown in **Circuit 2**.

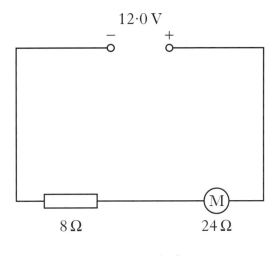

12·0 V

8 Ω   24 Ω

Circuit 2

State how this affects the speed of the motor compared to **Circuit 1**.

Explain your answer.   **2**

                                                                                        (10)

[Turn over

Marks

27. A mains operated mobile phone charger contains a transformer.

    Part of the circuit is shown below.

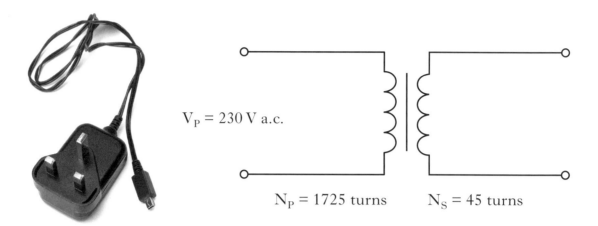

$V_P = 230$ V a.c.

$N_P = 1725$ turns $N_S = 45$ turns

The primary coil of the transformer has 1725 turns.

The secondary coil has 45 turns.

(a) Calculate the voltage across the secondary coil. 2

(b) When the charger is connected to a mobile phone the output current is 0·80 A. Calculate the current in the primary coil. 2

(c) What is the frequency of the mains supply in the UK? 1

(d) 230 V a.c. is the quoted value of the mains supply in the UK.

    State how the quoted value compares with the peak value. 1

    (6)

Marks

28. A photographic darkroom has a buzzer that sounds when the light level in the room is too high. The circuit diagram for the buzzer system is shown below.

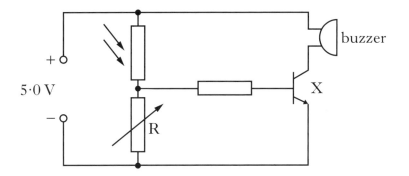

(a) (i) Name component X.   1

     (ii) What is the purpose of component X in the circuit?   1

(b) The darkroom door is opened and the light level increases.

     Explain how the circuit operates to sound the buzzer.   3

(c) The table shows how the resistance of the LDR varies with light level.

Light level (units)	LDR Resistance (Ω)
20	4500
50	3500
80	2500

The variable resistor has a resistance of 570 Ω.

The light level increases to 80 units.

Calculate the current in the LDR.   3

(d) What is the purpose of the variable resistor R in this circuit?   1

                                                                                           (9)

[Turn over

Marks

29. A lighthouse uses a converging lens to produce a beam of light.

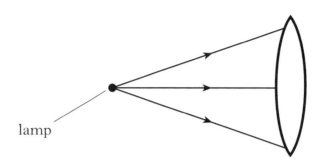

lamp

(a) The lamp is placed at the focal point of the lens.

Copy and complete the diagram to show the paths of the light rays after they pass through the lens.   **1**

(b) The power of the lens is 6·25 D.

Calculate its focal length.   **2**

(c) The lamp flashes once every 7·5 seconds.

What is the name given to the time between each flash?   **1**

(d) The lighthouse also uses a foghorn to alert ships.

A ship is at a distance of 2·04 km from the lighthouse.

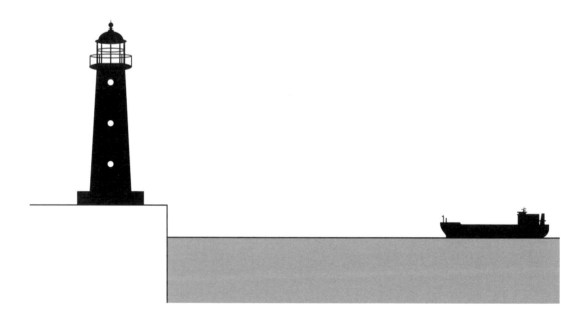

Calculate the time taken for the sound to reach the ship.   **2**

(e) Light waves are transverse waves. Sound waves are longitudinal waves.

Describe each type of wave in terms of vibrations.   **2**

(8)

Marks

30. A hospital radiographer calculates the equivalent dose of radiation absorbed by a patient. This is done by multiplying the absorbed dose by a radiation weighting factor.

    (*a*) State what is meant by a radiation weighting factor.   **1**

    (*b*) During a scan of the patient's brain, the absorbed dose is measured as 1·5 mGy. The mass of the brain is 1·4 kg.

        Calculate the energy absorbed by the brain during the scan.   **2**

    (*c*) In another medical procedure, a radioactive chemical is injected into a patient.

        The chemical is prepared by the technician from a source which has an activity of 320 MBq.

        The source has a half-life of 6 hours.

        Calculate the activity of the source 18 hours later.   **2**

        (5)

[Turn over

Marks

31. (*a*) A student is researching information on nuclear reactors.

The following diagram is found on a website.

It illustrates a type of reaction that takes place in a reactor.

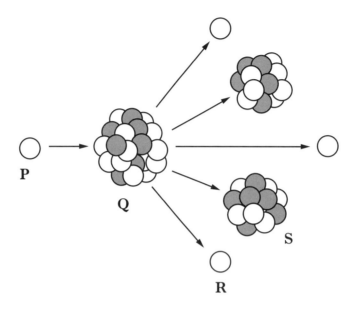

(i) What type of nuclear reaction is shown in the diagram? **1**

(ii) The labels have been omitted at positions **P**, **Q**, **R** and **S** on the diagram.

State clearly what each of these labels should be. **2**

(*b*) Name the part of the reactor whose function is to prevent release of radiation beyond the reactor. **1**

(*c*) Disposal of some types of radioactive waste from nuclear reactors is particularly difficult.

Give a reason for this difficulty. **1**

(*d*) Electricity can be generated using fossil fuels or nuclear fuel.

State one advantage of using nuclear fuel. **1**

(6)

[END OF QUESTION PAPER]

[BLANK PAGE]

Acknowledgements

Permission has been sought from all relevant copyright holders and Hodder Gibson is grateful for the use of the following:

Image © Darryl Brooks/Shutterstock.com (2013 page 11);
Image © chungking/Shutterstock.com (2013 page 18);
Image © design56/Shutterstock.com (2013 page 20).

INTERMEDIATE 2 | ANSWER SECTION

PHYSICS INTERMEDIATE 2
2009

SECTION A

1.	C	11.	B
2.	C	12.	D
3.	E	13.	A
4.	A	14.	D
5.	B	15.	C
6.	B	16.	E
7.	D	17.	E
8.	E	18.	A
9.	B	19.	D
10.	E	20.	C

SECTION B

21. (a) $E_p = mgh$
$$= 2000 \times 10 \times 540$$
$$= 10800000 \text{ J} \ (1 \cdot 08 \times 10^7 \text{ J})$$

(b) $E_k = \dfrac{1}{2}mv^2$
$$64000 = 0 \cdot 5 \times 2000 \times v^2$$
$$v^2 = 64$$
$$v = 8 \text{ m/s}$$

(c) (i) $P = IV$
$$45600 = I \times 380$$
$$I = 120 \text{ A}$$

(ii) $E = Pt$
$$= 45600 \times 60 \times 60$$
$$= 1 \cdot 64 \times 10^8 \text{ J}$$

22. (a) $a = \dfrac{(v-u)}{t}$

$$a = \dfrac{(3-0)}{5}$$

$$a = 0 \cdot 6 \text{ m/s}^2$$

(b) $F = ma$
$F = 40 \times 0 \cdot 6$
$$= 24 \text{ N}$$

(c) There is an unbalanced force/friction, which acts against the motion.

23. (a) width/length of card (d)
time taken for <u>card to cut beam</u> (t)
$$v = \dfrac{d}{t}$$
or
$$v = \dfrac{\text{length of card}}{\text{time taken for card to cut beam}}$$

(b) (i) $p = mv$
$$= (5 \times 10^{-4} + 0 \cdot 3) \times 0 \cdot 35$$
$$= 0 \cdot 105 \text{ kg m/s}$$

(ii) $v = \dfrac{p}{m} = \dfrac{0 \cdot 105}{5 \times 10^{-4}}$
$$= 210 \text{ m/s}$$

(c) (i) $a = \dfrac{(v-u)}{t}$

$$10 = \dfrac{(v-0)}{0 \cdot 2}$$
$$v = 2 \text{ m/s}$$

(ii) $d = \bar{v}t$
$$= 1 \times 0 \cdot 2$$
$$= 0 \cdot 2 \text{ m}$$

24. (a) $E_h = cm\Delta T$
$$= 4180 \times 0 \cdot 1 \times 15$$
$$= 6270 \text{ J}$$

(b) $E_h = ml$
$$= 0 \cdot 1 \times 3 \cdot 34 \times 10^5$$
$$= 3 \cdot 34 \times 10^4 \text{ J}$$

(c) (i) $33400 + 6270 = 39670 \text{ J}$

E		$= Pt$
39670	$= 125 \times t$
t	$= 317 \cdot (36) \text{ s}$

(ii) Heat energy will be gained from surroundings/other food etc.
Therefore more energy must be removed.

25. (a) (i)
I	$=$	$0 \cdot 075 \text{ A}$
V	$=$	IR
$4 \cdot 2$	$=$	$0 \cdot 075 \times R$
R	$=$	$56 \ \Omega$

(ii) stays the same
$$\dfrac{1 \cdot 3}{0 \cdot 023} = 56 \cdot 5 \qquad \dfrac{3 \cdot 6}{0 \cdot 064} = 56 \cdot 25$$

or as the voltage increases the current increases by the same ratio
or because it's a straight line <u>through the origin</u>

(b) (i) $R_t = R_1 + R_2$
$$= 270 + 390$$
$$= 660 \ \Omega$$

(ii) $\dfrac{1}{R_t} = \dfrac{1}{R_1} + \dfrac{1}{R_2}$
$$= \dfrac{1}{33} + \dfrac{1}{56}$$
$$= 0 \cdot 048$$
$R_t = 20 \cdot 76 \ \Omega$

26. (a) $\dfrac{I_p}{I_s} = \dfrac{V_s}{V_p}$

$$\dfrac{I_p}{I} = \dfrac{5}{230}$$
$$I_p = 0 \cdot 022 \text{ A}$$

(b) (i) $P = \dfrac{V^2}{R}$

$$10 = \dfrac{9^2}{R}$$
$$R = 8 \cdot 1 \ \Omega$$

(ii) $V_g = \dfrac{V_o}{V_i}$

$= \dfrac{9}{1 \cdot 5}$

$= 6$

(c) $Eff\% = \dfrac{P_o}{P_i} \times 100$

$= \dfrac{20}{25} \times 100$

$= 80\%$

27. (a) (i) short sight = the image is in focus in front of the retina

or

cannot see distant objects <u>clearly</u>

(ii) concave or diverging

(iii) $P = \dfrac{1}{f}$

$= (-)\dfrac{1}{0 \cdot 18}$

$= (-) 5 \cdot 6 \text{ D}$

(b) (i) refraction = the change in the speed or wavelength of light as it passes between two media (of different densities)

(ii) $v = f\lambda$

$3 \times 10^8 = f \times 7 \times 10^{-7}$

$f = 4 \cdot 29 \times 10^{14} \text{ Hz}$

$f = 4 \times 10^{14} \text{ Hz}$

(c) (i)

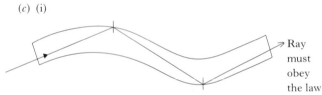

$\to$ Ray must obey the law

of reflection

Appropriate number of reflections

(ii) (total internal) reflection (TIR)

28. (a) $v = \dfrac{d}{t}$

$340 = \dfrac{d}{2 \times 10^{-3}}$

$d = 0 \cdot 68 \text{ m}$

$\therefore d = 0 \cdot 34 \text{ m one way}$

(b) $(f = \dfrac{1}{T})$

$f = \dfrac{1}{0 \cdot 125}$

$f = 8 \text{ Hz}$

(c) I = 200 mA

$P = IV$

$= 200 \times 10^{-3} \times 12$

$= 2 \cdot 4 \text{ W}$

(d) (i) (the resistor) stops too large a current (flowing <u>through</u> the LED) or too large a <u>voltage</u> <u>across</u> the LED

(ii) $V = 12 - 3 \cdot 5 = 8 \cdot 5$ (V)

$V = IR$

$8 \cdot 5 = 200 \times 10^{-3} \times R$

$R = 42 \cdot 5 \ \Omega$

29. (a) (i) <u>equipment</u>:

source, paper and suitable radiation detector and counter

<u>measurements</u>:

measure the count rate from each of the sources with paper and without paper between the source and the detector

<u>explanation</u>:

the source which produced a decreased count rate with paper is the alpha source

(ii) Cover the front window with a few mm of aluminium to stop beta.

(b)

Time	Activity
0	1
28	$\dfrac{1}{2}$
56	$\dfrac{1}{4}$
84	$\dfrac{1}{8}$
112	$\dfrac{1}{16}$

112 years

(c) (i) $H = Dw_R$

$= 20 \times 10^{-6} \times 20$

$= 400 \ \mu S_v$

or $400 \times 10^{-6} \ S_v$

(ii) increase distance (eg use tongs)

shielding (lead apron/gloves)

PHYSICS INTERMEDIATE 2 2010

SECTION A

1.	E	11.	B
2.	D	12.	B
3.	B	13.	C
4.	D	14.	A
5.	D	15.	A
6.	C	16.	D
7.	E	17.	C
8.	C	18.	E
9.	D	19.	A
10.	D	20.	E

SECTION B

21. (a) $a = \dfrac{v - u}{t}$

   $= \dfrac{6 - 0}{60}$

   $= 0 \cdot 1 \text{ m/s}^2$

(b) s = area under graph
   $= (0 \cdot 5 \times 60 \times 6) + (40 \times 6)$
   $= 420 \text{ m}$

(c) $v = \dfrac{s}{t}$

   $= \dfrac{420}{100}$

   $= 4 \cdot 2 \text{m/s}$

(d) $W = mg$
   $= 400 \times 10$
   $= 4000 \text{ N}$

(e) $F = ma$
   $= 400 \times 0 \cdot 1$
   $= 40 \text{ (N)}$

   Upward force $= 4000 + 40$
   $= 4040 \text{ N}$

22. (a) p before $= $ p after
   $(2 \cdot 0 \times 10^{-3} \times 4) = 3 \cdot 2 \times 10^{-3} \text{ v}$
   v $= 2 \cdot 5 \text{ m/s}$

(b)
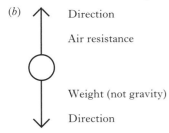
   Direction

   Air resistance

   Weight (not gravity)

   Direction

(c) $Q = It$

   $I = \dfrac{1650}{0 \cdot 15}$

   $= 1 \cdot 1 \times 10^4 \text{ A}$

(d) Light travels faster than sound

23. (a) $E_h = cm\Delta T$

   $c = \dfrac{2 \cdot 59 \times 10^7}{60 \times [\,(307 - (-173)\,]}$

   $= 899 \text{ J/kg}°\text{C}$

(b) $P = \dfrac{E}{t}$

   $t = \dfrac{2 \cdot 59 \times 10^7}{1440}$

   $= 18000 \text{ s}$

(c) $\dfrac{288000}{1440}$

   $= 200 \text{ (rocks)}$

(d) It would be easier
   Gravitational field strength at the surface of Mercury is less than that at the surface of Earth
   or
   Weight of rocks on Mercury is smaller than their weight on Earth
   or
   Gravity is less on Mercury

24. (a) $R_T = R_1 + R_2 = 8 + 24 = 32 \ \Omega$

   $V = IR$

   $I = \dfrac{6}{32}$

   $I = 0 \cdot 19 \text{ A}$

(b) $V_2 = \left(\dfrac{R_2}{R_1 + R_2}\right) V_S$

   $V_2 = \left(\dfrac{8}{8 + 24}\right) 6$

   $V_2 = 1 \cdot 5 \text{ V}$

   or

   $V = IR$
   $= 0 \cdot 19 \times 8$
   $= 1 \cdot 5 \text{ V}$

(c) Voltage across 8 Ω resistor would decrease
   The 8 Ω resistor now has a smaller proportion of the total resistance
   or less current in the circuit

25. (a) a.c. (source)

   <u>changing</u> magnetic field **or** <u>changing</u> current is necessary (to induce voltage)

(b) $P = IV$
   $= 0 \cdot 5 \times 12$
   $= 6 \text{W}$

(c) $P = IV$
   $= 0 \cdot 23 \times 23$
   $= 5 \cdot 3 \text{W}$

(d) percentage efficiency $= \dfrac{\text{useful } P_o}{Pi} \times 100$

   $= \dfrac{5 \cdot 3}{6} \times 100$

   $= 88 (\%)$

(e) $\dfrac{N_S}{N_P} = \dfrac{V_S}{V_P}$

   $V_S = \dfrac{3000 \times 12}{1500}$

   $= 24 \text{V}$

26. (a) Y (n-channel enhancement) MOSFET
   Z Lamp

(b) (Resistance) decreases

(c) (As resistance of thermistor decreases) voltage across thermistor decreases.
   V across X increases
   When it reaches 1·8V MOSFET $V_{(transistor)}$ switches on (Bulb lights and) buzzer sounds

(d) To allow switch on temperatures to be varied

27. (a) $s = vt$

   $t = \dfrac{51}{340}$

   $= 0 \cdot 15 \text{ s}$

(b) (i) Longitudinal
   (ii) A transverse wave is one in which the particles vibrate at right angles to the direction of the wave.
      A longitudinal wave is one in which the particles vibrate parallel to the direction of the wave.
   (iii) Energy

(c) (i) $P = \dfrac{V^2}{R}$

      $R = \dfrac{315^2}{2400}$

      $= 41\cdot34\ \Omega$

   (ii) *Any two from:*
- independent switching or one off/others stay on
- to ensure that 315 V is across each bulb
- if they were in series the necessary voltage would be too high

28. (a) (i) $v = f\lambda$

      $f = \dfrac{3 \times 10^8}{0\cdot06}$

      $= 5 \times 10^9\ Hz$

   (ii) $T = \dfrac{1}{f}$

      $T = \dfrac{1}{5 \times 10^9}$

      $= 2 \times 10^{-8}\ s$

(b) Signals received at same time
    Radio waves and microwaves have same speed

(c)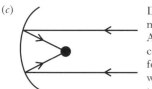
    Diagram must have minimum of two rays. All rays drawn must come to a distinct focus or incoming waves are focussed towards one point.

    Curved reflector gives <u>point where energy is maximised</u>

29. (a) A particle containing two protons and two neutrons
    or
    A helium nucleus

(b) The <u>gain/loss</u> of <u>electrons</u> by an <u>atom</u>

(c) $4800 \xrightarrow{1} 2400 \xrightarrow{2} 1200 \xrightarrow{3} 600 \xrightarrow{4} 300$ or equivalent

    $4 \times 2\cdot5 = 10$ hours

(d) $A = \dfrac{N}{t}$

    $N = 1200 \times 2 \times 60$
    $= 144\,000$ (decays)

(e) Source may also emit β and/or γ radiation

30. (a) (i) <u>Slows neutrons</u>
   (ii) <u>Absorbs neutrons</u>

(b) $P = \dfrac{E}{t}$

    $E = 1\cdot4 \times 10^9 \times 60 \times 60$
    $= 5\cdot0 \times 10^{12}$ (J)

Number of fissions $= \dfrac{5\cdot0 \times 10^{12}}{2\cdot9 \times 10^{-11}}$

    $= 1\cdot7 \times 10^{23}$

(c) Any valid advantage eg much greater energy per kg of fuel compared to other sources
    or
    No greenhouse gases emitted or equivalent

PHYSICS INTERMEDIATE 2 2011

SECTION A

1.	D	11.	D
2.	B	12.	B
3.	E	13.	B
4.	D	14.	D
5.	E	15.	E
6.	B	16.	C
7.	A	17.	C
8.	E	18.	D
9.	B	19.	A
10.	A	20.	A

SECTION B

21. (a) $s = vt$

    $\therefore\ t = \dfrac{11}{20}$

    $= 0\cdot55\ s$

(b) $a = \dfrac{v - u}{t}$

    $\therefore\ v = 10 \times 0\cdot55$

    $= 5\cdot5\ m/s$

(c)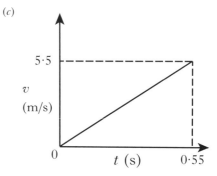

(d) s = area under graph **or** $s = \bar{v}t$

    $s = \dfrac{1}{2} \times 0\cdot55 \times 5\cdot5$ $s = \left(\dfrac{5\cdot5}{2}\right) \times 0\cdot55$

    $s = 1\cdot5\,m$ $s = 1\cdot5\,m$

22. (a) (i) Acceleration is the change of <u>velocity</u> (<u>not speed</u>) in <u>unit time</u>
   (ii) Direction of satellite is (continually) changing
      or
      <u>Velocity</u> of satellite is (continually) changing
      or
      There is an <u>unbalanced</u> (<u>not 'resultant'</u>) force on the satellite

(b) $F = 12 - 2 = 10N$
     $F = ma$
    $\therefore 10 = 50a$
     $a = 0\cdot2\,m/s^2$
    Direction is right

23. (a) $W = mg$
      $= 50,000 \times 10$
      $= 500,000N$

(b) $500,000N$

(c)   For scale drawing accept

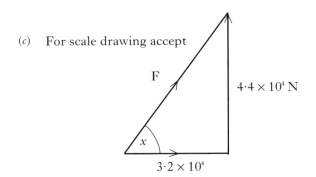

F^2 $= (3{\cdot}2 \times 10^4)^2 + (4{\cdot}4 \times 10^4)^2$
F $= 5{\cdot}4 \times 10^4 N$
$\tan x = \dfrac{4{\cdot}4 \times 10^4}{3{\cdot}2 \times 10^4}$
x $= 54°$
F $= 5{\cdot}4 \times 10^4 N$ at 036°

(d) $H = DW_R$
        $= 15 \times 10^{-6} \times 1$
        $= 1{\cdot}5 \times 10^{-5}$ Sv (15×10^{-6})

(e) Ionisation is when an <u>atom</u> gains or loses <u>electrons</u>

must have   one only needed

24. (a) (i) $(33 - 21) = 12°C$

(ii) $(120,000 - 12,000) = 108,000$ J

(iii) E_n $= cm\Delta T$
        $108,000 = c \times 2{\cdot}0 \times 12$
        c $= 4,500$ J/kg°C

(b) (i) Heat lost to <u>surroundings</u> (or similar)
        or water not evenly heated (or similar)
        Measured value of E_n too large **or** ΔT too small

(ii) Insulate beaker
        or Put lid on beaker
        or Stir water
        or Fully immerse heater

(c) E $= Pt$
        $108,000 = P \times 5 \times 60$
        P $= 360$ W

25. (a) $\dfrac{1}{R_T}$ $= \dfrac{1}{R_1} + \dfrac{1}{R_2}$

$= \dfrac{1}{4} + \dfrac{1}{2}$

$\therefore R_T = 1{\cdot}3\,\Omega$

(b) $R_T = R_1 + R_2$
        $= 1{\cdot}3 + 6$
        $= 7{\cdot}3\,\Omega$

(c) (Voltage across $2\,\Omega$ resistor = Voltage across $4\,\Omega$ resistor)
        V $= IR$
        $= 0{\cdot}1 \times 4$ (**or** $0{\cdot}2 \times 2$)
        $= 0{\cdot}4$ V

26. (a) dc – <u>electron</u> flows around a circuit in one direction only
        ac – <u>electrons'</u> direction changes/reverses continuously

(b) (i) ac **or** mains **or** one on left

(ii) Transformers are used to <u>change</u> the <u>magnitude</u> of an
        (alternating) voltage **or** current

(iii) Percentage efficiency $= \dfrac{useful\ P_o}{P_i} \times 100$

$useful\ P_o = \dfrac{30}{100} \times 50$

$=15$ W

27. (a) To reduce <u>current</u> in LED
        or
        To reduce <u>voltage</u> across LED
        or
        To reduce <u>power</u> to LED

(b) V $= 6 - 2 = 4\,V$
        V $= IR$
        $\therefore R = \dfrac{4}{0{\cdot}1}$
        $= 40\,\Omega$

(c) P $= I^2 R$
        $= (0{\cdot}1)^2 \times 40$
        $= 0{\cdot}4$ W

28. (a) (i) $P = IV$
        $= 0{\cdot}4 \times 10^{-3} \times 0{\cdot}5$
        $= 2 \times 10^{-4}$ W

(ii) $= \dfrac{4 \times 10^{-3}}{2 \times 10^{-4}}$

$= 20$ (cells)

(b) Light $\rightarrow$ electric (al)

(c) v $= f\lambda$

$\therefore \lambda = \dfrac{v}{f}$

$= \dfrac{3 \times 10^8}{6{\cdot}7 \times 10^{14}}$

$= 4{\cdot}5 \times 10^{-7}$m

29. (a) (i) P – Ultraviolet **or** UV
        Q – Microwaves

(ii) s $= vt$

$\therefore t = \dfrac{s}{v}$

$= \dfrac{4{\cdot}50 \times 10^{12}}{3 \times 10^8}$

$= 1{\cdot}5 \times 10^4$s

(iii) Decreases

(b) Q $= It$

$\therefore I = \dfrac{Q}{t}$

$= \dfrac{360}{60}$

$= 6$A

30. (a) P $= \dfrac{1}{f}$

$= \dfrac{1}{0{\cdot}03}$

$= 33$ D

(b) object

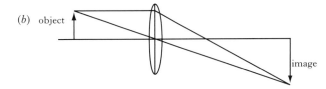

image

(c) Long sight
Converging lens brings light rays to <u>focus on retina</u> by <u>reducing focal length</u> (or equivalent).

31. (a) $N = At$

$= \dfrac{300 \times 10^{-6} \times 24 \times 60 \times 60}{}$

$= 26$ (decays)

(b) $2400 \to 1200 \to 600 \to 300$
$3 \times 5{,}730 = 17{,}190$ years

(c) An electron

(d) A helium <u>nucleus</u> **or** equivalent eg 2p + 2n

(e) Greater

(f) (i) (Aluminium) would stop <u>α particles</u> also

(ii) *Any two from:*
Shielding/Short times/Point away from people/
Increased distance/Wash hands

SECTION A

1.	D	11.	D
2.	A	12.	A
3.	D	13.	B
4.	B	14.	D
5.	C	15.	D
6.	A	16.	E
7.	C	17.	C
8.	B	18.	C
9.	C	19.	E
10.	A	20.	E

21. (a) (i) $d = vt$

$8300 \times 100 \times 60$
$= 49\,800\,000$ m

(ii) (As orbit is circular) <u>direction changes</u> / **or** <u>unbalanced force</u> exists
so <u>velocity changes</u>.

(b) $d = vt$
$800 \times 1000 = 300\,000\,000\,t$
$t = 0.0027$ s

(c) (i) The weight of 1 kg **or** Weight per unit mass **or** Earth's pull per kg.

(ii) 7.8 N/kg

(iii) $W = mg$
$= 84 \times 7.8$
$= 660$ N

22. (a) Car continues at a <u>constant speed</u> during this time.
AB represents driver's reaction time **or** the forces are balanced (or equivalent).

(b) $E = \dfrac{1}{2} mv^2$

$= 0.5 \times 700 \times 30^2$
$= 315\,000$ J

(c) 315 000 J

(d) $a = \dfrac{v - u}{t}$

$= (0 - 30)/2.5$
$(-)12(\text{m/s}^2)$

$F = ma$
$= 700 \times 12$
$= 8400$ N

or

$d =$ area under graph
$= 0.5 \times 2.5 \times 30$
$= 37.5$ (m)

$E_W = Fd$
$315\,000 = F \times 37.5$
$F = 8400$ N

23. (a) (i) $E_P = mgh$
$= 0.5 \times 10 \times 19.3$
$= 96.5$ J

(ii) $E_H = cm\Delta T$
$96.5 = 386 \times 0.50 \times \Delta T$
$\Delta T = 0.5°$ C

(iii) Less than.

Some heat is lost to surroundings/or equivalent.

(b) $E_h = ml$

$= 0.50 \times (2.05 \times 10^5)$

$= 102\,500$ J

24. (a)

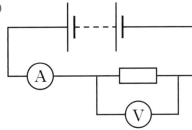

(b) $V = IR$

$5.7 = 0.60 \times R$

$R = 9.5\ \Omega$

(c) $P = VI$

$P = 5.7 \times 0.60$

$P = 3.42$ W

This is greater than the 3W or labelled power rating (so it overheats).

(d) No

In parallel the voltage is still the same/6V across each resistor

So power is the same

25. (a) MOSFET

(b) (Voltage) falls/decreases

(c) (i) $12 - 2.4 = 9.6$ V

(ii) $\dfrac{V_1}{V_2} = \dfrac{R_1}{R_2}$

$\dfrac{9.6}{2.4} = \dfrac{5600}{R_2}$

$R_2 = 1400\ \Omega$

(d) (Lamp) stays on

(Temperature falls)

R_T rises

V_T rises

$V_T > 2.4$ V **or** switching voltage

26. (a) (i) Speed of sound (much) less than speed of light (or similar)

(ii) $d = \bar{v}t$

$176 = \bar{v} \times 0.5$

$\bar{v} = 352$ m/s

(b) The current creates a magnetic field around the coil

The steel contact is attracted by the (magnetised) coil

(c) $V_R = 9 - 6 = 3$ V

$V = IR$

$3 = 800 \times 10^{-3} \times R$

$R = 3.75\ \Omega$

(d) $v = f\lambda$

$340 = 850 \times \lambda$

$\lambda = 0.4$ m

27. (a) Greater

(b)

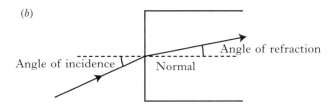

(c) Total internal reflection

28. (a)

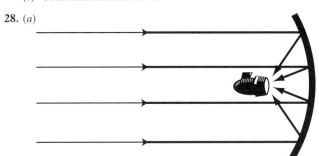

(b) More energy **or** power **or** amplitude is received (at the microphone.)

(c) $V_{gain} = \dfrac{V_o}{V_i}$

$V_{gain} = \dfrac{3}{0.024}$

$V_{gain} = 125$

(d) (i) Short sight

(ii) Diverging/concave lens

(e) $P = \dfrac{1}{f}$

$10 = \dfrac{1}{f}$

$f = 0.1$ m (10 cm)

29. (a) Half-life = 2 hours

(b) Any two valid answers.

(c) A type of <u>electromagnetic</u> radiation/wave/ray.

30. (a) (i) $D = E / m$

$= 0.000006/0.50$

$= 0.000012$ Gy

(ii) $H = Dw_R$

$= 0.000012 \times 20$

$= 0.00024$ Sv

(iii) $A = N / t$

$= 24{,}000/(5 \times 60)$

$= 80$ Bq

(b) (i) The moderator <u>slows neutrons</u>.

(ii) The containment vessel prevents/reduces release of radiations **or** radioactive gases **or** radioactive substances etc.

(c) Fission or Chain reaction.

PHYSICS INTERMEDIATE 2
2013

SECTION A

1.	B	11.	B
2.	D	12.	C
3.	E	13.	B
4.	D	14.	E
5.	E	15.	D
6.	A	16.	B
7.	D	17.	C
8.	A	18.	E
9.	A	19.	A
10.	C	20.	C

21. (a) $F = ma$

$$4500 = 750 \times a$$

$$a = 6 \text{ m/s}^2$$

(b) $a = \dfrac{v - u}{t}$

$$6 = \dfrac{54 - 0}{t}$$

$$t = 54 \div 6$$
$$t = 9 \text{ s}$$

(c) Other forces will act on the plane (e.g. drag)
or
Mass decrease (fuel consumption)

22. (a) 340 m/s

(b) (i) (Constant) <u>negative</u> acceleration

(ii) Distance = a.u.g.

$$= (0.05 \times 0.4) + (0.5 \times 0.05 \times 0.2)$$

$$= 0.02 + 0.005$$

$$= 0.025$$

$$= 0.025 \text{ m}$$

(iii) (Total) momentum before $= 0.50 \times 0.60 = 0.30$

(Total) momentum after $= 0.75 \times 0.40 = 0.30$

(iv) (A) $E_k = \tfrac{1}{2} mv^2$

Before $= \tfrac{1}{2} \times 0.5 \times 0.6^2 = 0.09$

After $= \tfrac{1}{2} \times 0.75 \times 0.4^2 = 0.06$

Loss $= 0.09 - 0.06$

$$= 0.03 \text{ J}$$

(B) Turns into <u>heat</u> energy (in pin/cork)

23. (a) (i) $d = vt$

$$= 2 \times 0.75$$

$$= 1.5 \text{ m}$$

(ii) $a = \dfrac{v - u}{t}$

$$10 = \dfrac{v - 0}{0.75}$$

$$v = 7.5 \text{ m/s}$$

(b) Same

All objects fall with the same (vertical) acceleration.

24. (a) $E_W = Fd$

$$= 250 \times 4.5$$

$$= 1125 \text{ J}$$

(b) $E_P = mgh$

$$= 144 \times 10 \times 0.75$$

$$= 1080 \text{ J}$$

(c) percentage efficiency $= \dfrac{\text{useful } E_o}{E_i} \times 100$

$$= \dfrac{1080}{1125} \times 100$$

$$= 96\%$$

25. (a) The energy given to the charges in a circuit.

(b) (i) $I = P/V$

$$= 1196/230$$

$$= 5.2 \text{ A}$$

(ii) $Q = I \times t$

$$= 5.2 \times (60 + 30)$$

$$= 468 \text{ C}$$

(iii) $E = mc\Delta T$

$$48000 = 0.25 \times 3900 \times \Delta T$$

$$\Delta T = 49.2° \text{ C}$$

(c) $\lambda = v/f$

$$= 3 \times 10^8 / 2500 \times 10^6$$

$$= 0.12 \text{ m}$$

26. (a) Lamp A

It has the lowest resistance/highest current/greatest power

(b) $P = V^2/R$

$$= 24^2/2.5$$

$$= 230 \text{ W}$$

(c)

(d) (i) 12 V

(ii) $1/R_p = 1/R_1 + 1/R_2$

$$= 1/8 + 1/24$$

$$= 4/24$$

$$R_p = 24/4$$

$$= 6 \ \Omega$$

(e) *Any one from:*
• The motor speed will reduce
• The (combined) resistance (of the circuit) is now higher/current is lower.
• Voltage across motor is less
• Motor has less power

27. (a) $N_s/N_p = V_s/V_p$

$$45/1725 = V_s/230$$

$$V_s = 6 \text{ V}$$

(b) $I_p V_p \quad = I_s V_s$

$\quad I_p \times 230 \quad = 0{\cdot}80 \times 6$

$\qquad I_p = (0{\cdot}80 \times 6)/230$

$\qquad\quad = 0{\cdot}021$ A or 21 mA

(c) 50 Hz

(d) The quoted value is smaller than the peak value.

28. (a) (i)(NPN) transistor

    (ii) To act as a <u>switch</u>

(b) Resistance of <u>LDR</u> reduces so voltage across <u>LDR</u> reduces
Voltage across <u>variable resistor/R</u> increases

When voltage across <u>variable resistor/R</u> is high enough
(0·7V) the transistor switches buzzer on.

(c) 80 units: resistance of LDR = 2500 (Ω)
Total resistance = 2500 + 570
$\qquad\qquad\qquad = 3070\ (\Omega)$

$I \quad = V/R$

$\quad = 5/3070$

$\quad = 1{\cdot}63 \times 10^{-3}$ A or 1·63 mA

(d) The variable resistor is to set the light level at which the
transistor will switch on or to set the level at which the
buzzer will sound.

29. (a)

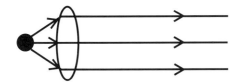

(b) focal length = 1/lens power
$\qquad\qquad\quad = 1/6{\cdot}25$
$\qquad\qquad\quad = 0{\cdot}16$ m

(c) The period

(d) $\quad d = v \times t$
$\quad 2040 = 340 \times t$
$\qquad t = 2040/340$
$\qquad\quad = 6$ s

(e) With <u>transverse</u> waves the vibrations are at <u>right angles</u> to
the <u>direction</u> of travel.

With <u>longitudinal</u> waves the vibrations are in the <u>same
direction</u> of travel.

30. (a) A measure of the biological effect of a radiation.

(b) $D = \dfrac{E}{m}$

$1{\cdot}5 \times 10^{-3} = \dfrac{E}{1.4}$

$E = 2{\cdot}1 \times 10^{-3}$ J

(c) 18 hours = 3 half lives
$320 \xrightarrow{\ D\ } 160 \rightarrow 80 - 40$ MBq
Activity after 18h = 40 MBq

31. (a) (i) Fission

    (ii)P (slow) neutron
       Q (fissionable) nucleus
       R (fast) neutron
       S fission fragment/daughter product

(b) Containment vessel

(c) Stays (highly) radioactive for a (very) <u>long time</u>

(d) Any valid answer eg Much more energy per kg of fuel
nucleus
or
Does not produce greenhouse/acidic gases